ATOMIC CHARGES, BOND PROPERTIES, AND MOLECULAR ENERGIES

ATOMIC CHARGES, BOND PROPERTIES, AND MOLECULAR ENERGIES

SÁNDOR FLISZÁR

**WILEY**

A JOHN WILEY & SONS, INC., PUBLICATION

Published by John Wiley & Sons, Inc., Hoboken, New Jersey
Published simultaneously in Canada

For general information on our other products and services or for technical support, please contact our
Customer Care Department within the United States at (800) 762-2974, outside the United States at
(317) 572-3993 or fax (317) 572-4002.

Wiley also publishes its books in variety of electronic formats. Some content that appears in print may
not be available in electronic formats. For more information about Wiley products, visit our web site
at www.wiley.com.

Library of Congress Cataloging-in-Publication Data:

Fliszár, Sándor.
    Atomic charges, bond properties, and molecular energies / Sándor Fliszár.
        p. cm.
    Includes index.
    ISBN 978-0-470-37622-5 (cloth)
1.   Chemical bonds.   I. Title.
    QD461.F536 2009
    541′.224--dc22

                                                                                            2008021431

Printed in the United States of America
10   9   8   7   6   5   4   3   2   1

CONTENTS

PREFACE   IX

DEDICATION AND ACKNOWLEDGMENTS   XI

I   CHARGE DISTRIBUTIONS   1

1   INTRODUCTION   3

    1.1   The Bond Energy Model / 3
    1.2   Scope / 5

2   THEORETICAL BACKGROUND   9

    2.1   The Hartree–Fock Approximation / 9
    2.2   Hartree–Fock–Roothaan Orbitals / 12
    2.3   Configuration Interaction Calculations / 12

3   CORE AND VALENCE ELECTRONS   17

    3.1   Introduction / 17
    3.2   Atomic Core and Valence Regions / 18
    3.3   The Valence Region Energy of Atoms / 25
    3.4   Summary / 34

4 THE VALENCE REGION OF MOLECULES   35

4.1   Model / 35
4.2   The Core–Valence Separation in Real Space / 36
4.3   Formula for the Valence Region Energy / 40
4.4   Interface with the Orbital Model / 43
4.5   Approximation for the Valence Energy / 46
4.6   Perturbation of the Valence Region / 49
4.7   Summary / 50

5 INDUCTIVE EFFECTS; ATOMIC CHARGES   53

5.1   Introduction / 53
5.2   The Inductive Effects / 54
5.3   Meaningful Atomic Charges / 58
5.4   Selected Reference Net Atomic Charges / 61

6 ATOMIC CHARGES AND NMR SHIFTS   65

6.1   Scope / 65
6.2   Introduction / 66
6.3   Merits of Charge–Shift Relationships / 66
6.4   Aromatic Hydrocarbons / 68
6.5   Relationships Involving sp^3 Carbon Atoms / 72
6.6   Relationships Involving Olefinic Carbons / 74
6.7   Carbon Bonded to Nitrogen or Oxygen / 76
6.8   Correlations Involving N-15 NMR Shifts / 77
6.9   Correlations Involving O-17 Atoms / 83
6.10 Summary / 86

7 CHARGES AND IONIZATION POTENTIALS   89

7.1   Conclusion / 91

8 POPULATION ANALYSIS   93

8.1   The Standard Mulliken Formula / 93
8.2   Modified Population Analysis / 94
8.3   An Adequate Approximation / 95
8.4   Conclusions / 97

II   CHEMICAL BONDS: ENERGY CALCULATIONS   99

9   THERMOCHEMICAL FORMULAS   101

9.1   Basic Formulas / 101

9.2   Zero-Point and Heat Content Energies / 103

9.3   Concluding Remarks / 110

10   THE CHEMICAL BOND: THEORY (I)   113

10.1   Synopsis / 113

10.2   Nonbonded Interactions / 115

10.3   Reference Bonds / 117

10.4   Bond Energy: Working Formulas / 120

10.5   Basic Theoretical Parameters / 126

10.6   Saturated Molecules / 128

11   THE CHEMICAL BOND: THEORY (II)   133

11.1   Valence Atomic Orbital Centroids / 134

11.2   Unsaturated Systems / 141

11.3   Recapitulation / 148

12   BOND DISSOCIATION ENERGIES   151

12.1   Scope / 151

12.2   Theory / 152

12.3   Nonbonded Interactions / 156

12.4   Selected Reorganizational Energies / 157

12.5   Applications / 158

12.6   Conclusion / 166

III   APPLICATIONS   167

13   SATURATED HYDROCARBONS   169

13.1   Acyclic Alkanes / 169

13.2   Cycloalkanes / 171

14 UNSATURATED HYDROCARBONS   177

14.1   Olefins / 177
14.2   Aromatic Molecules / 182

15 NITROGEN-CONTAINING MOLECULES   189

15.1   Amines: Charges of the Carbon Atoms / 189
15.2   Nitrogen Charges and Bond Energies / 190
15.3   Results / 194

16 OXYGEN-CONTAINING MOLECULES   197

16.1   Ethers / 197
16.2   Alcohols / 199
16.3   Carbonyl Compounds / 201

17 PERSPECTIVES   205

APPENDIX: WORKING FORMULAS   207

A.1   Charge–NMR Shift Correlations / 207
A.2   General Energy Formulas / 209
A.3   Bond Energy Formulas / 213

BIBLIOGRAPHY   219

INDEX   231

PREFACE

This book is about atomic charges, chemical bonds, and bond energy additivity. However, nuclear magnetic resonance, inductive effects, zero-point and heat content energies, and other topics are an integral part of this study, to achieve our goals.

The electronic charges, the bond energies, and—way down the line—the energy of atomization and the enthalpy of formation of organic molecules are what we shall calculate with chemical accuracy, in a simple manner.

Of course, new ideas have to be implemented, in terms of both bond energies and the calculation of atomic charges; the formula describing the energy of a chemical bond in a ground-state molecule—the "intrinsic" bond energy—translates intuitive expectations, namely, that the energy of a bond formed by atoms k and l should depend on the amount of electronic charge carried by these atoms. The all-important relationship between the *intrinsic bond energies*—which apply for bonds in molecules at equilibrium—and the corresponding familiar *bond dissociation energies*—which refer specifically to the process of bond breaking—is also described.

Intrinsic bond energies and bond dissociation energies meet different practical needs. The former play an important role in the description of ground-state molecules. Dissociation energies, on the other hand, come into play when molecules undergo reactions. Now, any interaction between a molecule and its environment (such as complex formation or the adsorption onto a metallic surface, or hydrogen bonding, for example) affects its electron distribution and thus the energies of its chemical bonds. If we figure out the relationship between dissociation and intrinsic bond energies, we could begin to understand how the environment of a molecule can promote or retard the dissociation of one or another bond of particular interest in that molecule. This outlook hints at a rich potential of future research exploiting charge analyses to gain insight into *local* molecular properties, specifically, into bond energies, first, and, going from there, into matters of great import regarding the dissociation of chemical bonds.

Our methods, like many others, do not answer all possible questions; plain quantum chemistry is undeniably the procedure of choice. But what our methods do, they surely do it with great accuracy and with a chemical insight that is not offered by traditional approaches and, moreover, in an extremely inexpensive manner, in regard to computational resources and costs. Applications are presented for saturated acyclic and cyclic hydrocarbons, amines, alcohols, ethers, aldehydes and ketones, and ethylenic and aromatic hydrocarbons. The results surely are impressive. Beyond offering new perspectives to old problems, the wealth of details given here hopefully lays fertile grounds for future breakthroughs along the simple lines advocated in this work. And while keeping the text as accessible as possible, the bibliography is richly designed, to assist interested readers in going more and more deeply into the wheelwork of the matter. With that goal in mind, a compendium of user-friendly final formulas is offered in the Appendix.

A few words about this book are in order. It is about chemistry (or, should I say physical organic chemistry?) that exploits quantum-mechanical methods. First-year graduate and advanced undergraduate introductory courses in quantum chemistry offer the required background. A most concise presentation—not an explanation— of the required quantum-mechanical techniques is offered, but a large part of this book is accessible without mastering them. The physics are given with sufficient details, which, I hope, are easy to follow. The whole thing—hopefully with the future aid of interested readers—adds powerful new investigative tools in areas of great import that help the reader understand chemical principles and predict properties.

This is the present for the future.

SÁNDOR FLISZÁR

DEDICATION AND ACKNOWLEDGMENTS

I wish to dedicate this work to my students and postdoctoral fellows who made things possible: Marie-Thérèse Béraldin, Jacques Bridet, Jean-Louis Cantara, Guy Cardinal, Michel Comeau, Geneviève Dancausse, Normand Desmarais, Anikó Foti, Marielle Foucrault, Annik Goursot, Jacques Grignon, Hervé Henry, Gérard Kean, Claude Mijoule, Camilla Minichino, Andrea Peluso, François Poliquin, Réal Roberge, and Édouard C. Vauthier. I also express my heartfelt thanks for active help, advice, patience, and friendship to Professors Giuseppe Del Re, Vincenzo Barone, Jean-Marie Leclercq, Simone Odiot, and Dennis Salahub. But my special thanks go to Steve Chrétien for his great dedication in helping me with most of the numerical calculations rooted in Del Re's theory that support the present work. It was my fortune to have him as a coworker and, I think, as a friend. And had it not been for the patient and skilled help of Édouard Vauthier, and his friendship, many of these pages could not have been written. I am also indebted to my daughter Saskia for helpful suggestions and corrections to this text. I also wish to thank Mr. Xiao-Gang Wang for his kindness and skillful help. Last but not least, I include Signora Dora and Don Gaetano Lampo in this dedication. These fine people made me a better person by teaching me important things, such as the *true* tolerance that is so uniquely part of Neapolitan culture. They also made me a little, but just a little, fatter.

Now that everything is behind us, I begin to understand how we benefited from exceptional circumstances when most important scientific facts were harvested with an incredible timing; the ink had barely dried when they came to fill essential needs in the fabric of our work. We could not have done it otherwise. I feel deeply in debt and wish to express my sincere appreciation to my fellow chemists and physicists recalling the wording of an old chinese proverb: *When you drink the water, don't forget the source.* Thank you. I won't forget.

SÁNDOR FLISZÁR

PART I

CHARGE DISTRIBUTIONS

CHAPTER 1

INTRODUCTION

1.1 THE BOND ENERGY MODEL

This book is about electronic charge distributions, chemical bonds, bond energy additivity in organic molecules, and the description of their relevant thermochemical properties, such as the energy of atomization, the enthalpy of formation, and the like, using computer-friendly methods.

Additivity schemes with fixed bond energy (or enthalpy) parameters plus a host of corrective factors reflecting nonbonded steric interactions have a long history in the prediction of thermochemical properties, such as the classical enthalpy of formation of organic molecules. Allen-type methods, for example, nicely illustrate the usefulness of empirical bond additivity approaches [1,2].

But theory tells a different story.

Immutable bond energy terms tacitly imply never-changing internuclear distances between atoms whose electron populations would never change. But the point is that invariable local electron populations cannot describe a set of electroneutral molecules. Unless the net atomic charges of all atoms in all molecules always exactly equal zero, any additivity scheme postulating fixed atomic charges obviously violates all requirements of molecular electroneutrality; for example, if the same carbon net charges and the same hydrogen net charges ($\neq 0$) are assigned to the carbon and hydrogen atoms of methane and ethane, simple charge normalization indicates that molecular electroneutrality cannot be satisfied for both molecules. Finally, unless we stipulate that atomic charges have no bearing on bond energies or else, that

Atomic Charges, Bond Properties, and Molecular Energies, by Sándor Fliszár
Copyright © 2009 John Wiley & Sons, Inc.

any change in atomic charge is perfectly counteracted by appropriate changes in internuclear distances in order to prevent changes of bond energy, we are led to the concept of bond energies depending on the charges of the bond-forming atoms.[1]

But quantum chemistry *hic et nunc* does not know about chemical bonds, unless we say so. The approach chosen here is centered on the potentials at the nuclei found in a molecule.

The Hellmann–Feynman theorem tells us that all forces in a molecule can be understood on purely classical grounds, provided that the exact electron density (or at least a density derived from a wave function satisfying this theorem) is known for that molecule [3]. We have exploited this vein. The description of atomization energies, on the one hand, and that of bond energies, on the other, were reduced to purely electrostatic problems involving only nuclear–electronic and nuclear–nuclear interactions. There is a price to be paid for this simplification—the *mechanism* of bond formation cannot be understood in purely electrostatic terms [4,5]. The kinetic energy of the electrons plays a decisive role because the electronic Hamiltonian of an atom or a molecule is bounded from below only if the kinetic energy is duly accounted for. This decisive role deeply reflects the theory explaining why chemical bonds are formed in the first place [4,5]. In short, our electrostatic approach allows no inquiry into the origin of chemical bonds. In contrast, it is well suited for describing chemical bonds as they are found in molecules at equilibrium.

This is so because the Hellmann–Feynman theorem offers a most convenient way to bring out the main features of chemical binding. By taking the nuclear charges as parameters, the binding of each individual atom in a molecule can be defined without having recourse to an a priori real-space partitioning of that molecule into atomic sub-spaces. This binding is determined entirely by the potentials at the atomic nuclei. In short, our definition of bond energies does not involve virtual boundaries that delimit the space assigned to the individual atoms in a molecule with intent to subsequently describe the chemical bonds linking them. We know, of course, that powerful and realistic methods describe a useful partitioning of molecules into "atoms in a molecule" [6]—an approach that certainly offers much chemistry. The methods developped here simply represent another perspective of the same problem. Important arguments concern a real-space core–valence charge partitioning, a topic that resists the approach [6] leading to the concept of "atoms in the molecule." Moreover, our approach involves the use of Gauss' theorem and a sensible application of the Thomas–Fermi model [7,8], which is known to give reasonably accurate atomic energies with the use of Hartree–Fock densities [9–12].

However, direct calculations of accurate bond energies represent a major challenge. Examples are given [13,14] where the ratios of carbon–carbon bond energies, relative to that of ethane, were successfully calculated for ethylene, acetylene, benzene, and

---

[1]Empirical bond additivity methods circumvent the problems linked to charge normalization constraints because they modify the genuine "bond energies" by tacit inclusion of extra terms that have nothing to do with bond energy itself. Still, a number of additional "steric factors" must be introduced in order to achieve what fixed bond energy terms alone cannot do: an agreement with experimental results (see Chapter 10).

TABLE 1.1. CC Bond Energies Relative to That of Ethane

Basis[a]	Ratio of CC Bond Energies			
	C_2H_4	C_2H_2	C_6H_6	$c\text{-}C_3H_6$
1 6−31 G(d,p)	1.84_6	2.82_8	1.59_0	1.00_2
2 6−311 G(d,p)	1.93_8	3.04_8	1.60_1	0.99_3
3 6−311 G($2df,2p$)	1.95_5	2.90_8	1.60_3	0.97_2
4 6−311 G($2df,2pd$)	1.97_2	2.96_0	1.62_1	0.97_7
5 vD($2d,2p$) BLYP	1.99_1	3.09_4	1.61_5	0.95_1
6 vD($2d,2p$) B3LYP	1.91_5	2.98_0	1.58_1	0.95_8
7 vD($2df,2pd$) BLYP	2.02_0	3.15_4	1.63_4	0.95_9
8 vD($2df,2pd$) B3LYP	1.94_3	3.03_6	1.59_7	0.96_2
Semiempirical	2.000		1.640	

[a]The basis set of **1** is from Ref. 15, that of **2** from Ref. 16, and those of **3** and **4** are from Ref. 17. In **5−8** we used van Dujneveldt's bases [18], also adding a set of f functions on C and of d functions on H [17] (**7,8**), Becke's gradient correction to exchange [19], and the Lee−Yang−Parr (LYP) potential [20]. The B3LYP functional involves a fully coherent implementation, whereby the self-consistent field (SCF) process, the optimized geometry, and the analytic second derivatives are computed with the complete density functional, including gradient corrections and exchange.

cyclopropane (Table 1.1). For ethane itself, the calculated CC bond energy amounts to ~70 kcal/mol from Hartree−Fock calculations [13] or 68.3 ± 0.4 kcal/mol, as given by density functional calculations [14]. The best fit with experimental data is obtained with 69.633 kcal/mol for that bond, which is satisfactory.

1.2 SCOPE

The difficulties encountered in the direct calculation of bond energies can be overcome—with hard labor and some approximations—in only a few cases, but the good news is that only a few reference bond energies need to be calculated for model systems. Those determined for the CC and CH bonds of ethane, for example, are sufficient for the description of saturated hydrocarbons; the addition of the reference bond energy describing the double bond of ethene extends the range of applications to olefinic molecules, including polyenic material. It is thus well worth the trouble to calculate a few reference bond energies—and this can be done with reasonable accuracy—because the rest follows as explained here. That is where atomic charges come into the picture and solve the problem presented here.

The description of bond energies that depend explicitly on the charges of the bond-forming atoms is attractive for the concepts it applies and for its usefulness in the prediction of important thermochemical quantities, such as the energy of atomization or the enthalpy of formation of organic molecules. But its success critically depends on the availability of accurate charge results.

Now, the problems associated with the search for *meaningful* atomic charges, such as those required in applications of our bond energy formulas, are manifold. One concerns the selection of the population analysis. As is well known, Mulliken charges represent only one of the possible definitions of atomic net charges [21–29]; widely different values are produced with comparatively small changes in a basis set, even in ab initio SCF calculations. Since trends within similar series are generally little affected by the computational scheme, this has seldom received significant attention in applications to chemical problems. But the present situation is one where the choice of the basis set could play a very important role, as does inclusion of basis superposition effects [30]. While Mulliken's population analysis [31] is probably the most widely used one, Löwdin's method [32] or Jug's approach [33] could also be envisaged with success. Selection of the method does greatly affect the final numerical results, as vividly exemplified by the SCF net charges obtained with a minimal basis set for the nitrogen of methylamine, namely, -374, $+47$, and -291 millielectron (1 me $= 10^{-3}$ e units), respectively, depending on whether Mulliken's, Löwdin's, or Jug's definition is implemented. So, under these circumstances, things do not seem encouraging. Fortunately, they are not really as bad as suggested by these examples. We shall learn about charge *variations* suited for bond energy calculations that withstand comparison with experiment. A promising example is offered by the calculated charge variations Δq_N of the nitrogen atoms of a series of amines, relative to that of methylamine, as given by the Mulliken, Löwdin, and Jug methods (Table 1.2). The examples shown in Table 1.2 are meant to illustrate how different ways of partitioning overlap populations do affect calculated atomic charge variations accompanying structural modifications; admittedly, the differences can be relatively minor, but the absolute values of the same charges defy any reasonable expectation. Evidently, one should not rely too heavily on numbers calculated by population analysis. Mulliken's assignment of half the overlap probability density to each atomic orbital (AO) is rather arbitrary and occasionally leads to unphysical results. Still, we shall see that the Mulliken scheme offers a valid starting point. To get useful charges, however, we must rethink the problem of assigning overlap populations—a topic highlighted in Chapter 8.

At first, it may seem surprising that many methods give results in semiquantitative agreement with Mulliken population analysis values, but it now appears that it is rather what one should expect from any sensible method. This holds true for the simplest

TABLE 1.2. Calculation of Δq_N in Amines

Molecule	Δq_N (me)		
	Mulliken	Löwdin	Jug
CH_3NH_2	0	0	0
$C_2H_5NH_2$	-5.16	-4.86	-5.01
$n\text{-}C_3H_7NH_2$	-4.96	-4.65	-4.86
$iso\text{-}C_3H_7NH_2$	-9.00	-8.90	-9.20
$iso\text{-}C_4H_9NH_2$	-3.59	-3.84	-4.10
$tert\text{-}C_4H_9NH_2$	-11.58	-11.94	-12.49

possible one [29]: Del Re's approach is based on rough semiempirical approximations of simple molecular orbital – linear combination of atomic orbitals (MO-LCAO) theory of localized bonds. The original, extremely simple parameterization that reproduced electric dipole moments [29] and a more recent one [34–36] for charges that correlate with nuclear magnetic resonance (NMR) shifts and that are fit for accurate energy calculations, turn out to correspond to solutions of MO-LCAO Mulliken-type population analyses differing from one another by the mode of partitioning overlap terms [37]. The link between Del Re's simple semiempirical approach and the more familiar MO-LCAO charge analyses is clearly established.

While the charges for use in our description of bond energies were originally obtained from accurate SCF computations using a variant of Mulliken's population analysis, the observation [38–44] that the ^{13}C, ^{15}N, and ^{17}O nuclear magnetic resonance (NMR) shifts are linearly related to these charges permitted rapid progress in the application of the charge-dependent bond energy formulas to thermochemical problems since NMR results are more readily available than are good-quality population analyses. Of course, this strategy presumes not only a justification for assumed correlations between NMR shifts and net atomic charges (which is described in Chapter 6) but also a solid knowledge of well-justified charge analyses. This is no minor task.

The original definition of atomic charges found in the CH backbone of organic molecules is rooted in the idea that the carbon charges vary as little as possible on structural modification. This has triggered *inductive reasoning*, which maintains that if a situation holds in all *observed* cases, then the situation holds in all cases. Indeed, detailed tests involving ^{13}C chemical shifts, the ionization potentials of selected alkanes, and, most importantly, thermochemical data, unmistakably point to identical sets of charge values. Now, of course, the problem of induction is one of considerable controversy (and importance) in the philosophy of science; we must thus be extremely cautious in attributing physical meaning to atomic charges. In the approach known as *instrumentalism*, one could as well consider them as convenient ideas, useful instruments to explain, predict, and control our experiences; the empirical method is there to do no more than show that theories are consistent with observation. With these ideas in mind, atomic charges that are now widely used in molecular dynamics calculations and in the evaluation of solvation energies within the generalized Born approach also deserve renewed attention.

Admittedly, things have not been easy. But now we can benefit from the beauty of simplicity and learn about one unique kind of atomic charge rooted in quantum theory: the only ones that satisfy highly accurate correlations with experimental NMR shift results and that are at the same time directly applicable to bond energy calculations.

But the key to the theory of bond energy is in the description of real-space core and valence regions in atoms and molecules: therein lies the basic idea that gives rise to the notion of molecular chemical binding, expressed as a sum of atomlike terms. That marks the beginning of our story.

CHAPTER 2

THEORETICAL BACKGROUND

The electron, discovered by J. J. Thomson in 1895, was first considered as a corpuscule, a piece of matter with a mass and a charge. Nowadays things are viewed differently. We rather speak of a "wave–particle duality" whereby electrons exhibit a wavelike behavior. But, in Levine's own words [45], quantum mechanics does not say that an electron is distributed over a large region of space as a wave is distributed; it is the probability patterns (wavefunctions) used to describe the electron's motion that behave like waves and satisfy a wave equation.

Here we benefit from the notion of *stationary* electron density. The particles are not at rest, but the probability density does not change with time.

2.1 THE HARTREE–FOCK APPROXIMATION

The Hartree–Fock self-consistent field (SCF) method is the primary tool used in this chapter. It is rooted in the time-independent one-electron Schrödinger equation (in atomic units):

$$\left[-\frac{1}{2}\nabla^2(1) + V(r_1) \right] \phi_i(1) = \epsilon_i(1)\phi_i(1) \qquad (2.1)$$

The effective one-electron operator indicated in brackets includes the kinetic energy operator $-\frac{1}{2}\nabla^2$ and an effective potential energy $V(r_1)$ taken as an averaged function of r_1—the distance of electron 1 from the nucleus. In this approximation, electron 1

Atomic Charges, Bond Properties, and Molecular Energies, by Sándor Fliszár
Copyright © 2009 John Wiley & Sons, Inc.

moves in the field created by both the nuclear charge Z and a smeared-out static distribution of electric charge due to electrons $2, 3, \ldots, n$. The eigenfunction $\phi_i(1)$ is a one-electron orbital and $\epsilon_i(1)$ is the corresponding energy.

The Hartree–Fock equation

$$\hat{F}\phi_i = \epsilon_i\phi_i \tag{2.2}$$

has the same form as (2.1) but introduces spin explicitly in the description of the wavefunction. The ϕ_i terms are now spin orbitals, and ϵ_i is the eigenvalue of spin orbital i. The effective Hartree–Fock Hamiltonian $\hat{F}$ contains 2 one-electron operators, namely, the kinetic energy operator

$$\hat{T} = -\frac{1}{2}\nabla^2 \tag{2.3}$$

and the potential energy for the attraction between the electron and the nucleus of charge Z:

$$\hat{V}_{ne} = -\frac{Z}{r} \tag{2.4}$$

In addition, $\hat{F}$ contains two bielectronic operators. They describe the interaction between the electron occupying spin orbital i and the other electrons found in the atom. So, for the interaction between electrons 1 and 2 at a distance r_{12}, we have the Coulomb operator $\hat{J}_j$ and the exchange operator $\hat{K}_j$ defined by

$$\hat{J}_j(1)\phi_i(1) = \phi_i(1)\int |\phi_j(2)|^2\frac{1}{r_{12}}d\tau_2 \tag{2.5}$$

$$\hat{K}_j(1)\phi_i(1) = \phi_j(1)\int \frac{\phi_j^*(2)\phi_i(2)}{r_{12}}d\tau_2 \tag{2.6}$$

where $d\tau$ is the volume element

$$d\tau = r^2\sin\theta\,d\theta\,d\varphi\,dr \tag{2.7}$$

and the subscript 2 refers to electron 2. Of course, similar Coulomb and exchange operators for electrons $3, \ldots, n$ are also part of $\hat{F}$. The Coulomb integral $\int \phi_i^*(1)\hat{J}_j(1)\phi_i(1)\,d\tau_1$ represents the repulsion between electron 1 and a smeared-out electron with density $|\phi_j(2)|^2$. The exchange integral $\int \phi_i^*(1)\hat{K}_j(1)\phi_i(1)\,d\tau_1$ arises from the requirement that the wavefunction be antisymmetric with respect to electron exchange in order to satisfy Pauli's indistinguishability principle of identical particles. Because of the occurrence of ϕ_j terms in the Coulomb and exchange operators and, thus, in $\hat{F}$, Eq. (2.2) must be solved iteratively until self-consistency is achieved, resulting in a set of self-consistent field (SCF) eigenfunctions $\phi_{1s}, \phi_{2s}, \ldots$, with orbital eigenvalues $\epsilon_{1s}, \epsilon_{2s}, \ldots$

These Hartree–Fock energies are reasonably good approximations to the orbital energies of an atom, as determined by X-ray and optical spectroscopy term values. We take them as the energies of the individual electrons, knowing, of course, that each of these orbital energies is computed over the entire atom and that no localization of the individual electrons in certain regions of space should be attempted. The energy ϵ_i of an electron in orbital ϕ_i consists of its kinetic energy

$$T_i = \int \phi_i^* \hat{T} \phi_i \, d\tau \tag{2.8}$$

and of its potential energy. The interaction between that electron and the nucleus is as follows, from Eq. (2.4):

$$V_{\text{ne},i} = -Z \int \frac{|\phi_i(r)|^2}{r} \, d\tau \tag{2.9}$$

On the other hand, the interaction between that electron, denoted here as electron 1, and electrons 2, 3, ..., n is given by the appropriate sums of Coulomb and exchange integrals, for example, for electron 2 interacting with electron 1:

$$J_{ij} = \int \int |\phi_i(1)|^2 \frac{1}{r_{12}} |\phi_j(2)|^2 \, d\tau_1 \, d\tau_2 \tag{2.10}$$

$$K_{ij} = \int \int \frac{\phi_i^*(1)\phi_j(1)\phi_j^*(2)\phi_i(2)}{r_{12}} \, d\tau_1 \, d\tau_2 \tag{2.11}$$

These integrals are most conveniently carried out using the following expansion [46] of $1/r_{12}$ in terms of spherical harmonics

$$\frac{1}{r_{12}} = \sum_{l=0}^{\infty} \sum_{m=-l}^{l} \frac{4\pi}{2l+1} \cdot \frac{r_<^l}{r_>^{l+1}} [Y_l^m(\theta_i, \varphi_i)]^* Y_l^m(\theta_j, \varphi_j) \tag{2.12}$$

where $r_>$ is the larger and $r_<$ is the smaller of r_1 and r_2. The total repulsive potential energy experienced by electron 1 in orbital ϕ_i is thus computed from a sum of Coulomb integrals, $J_{12} + J_{13} + \cdots$, from which we subtract the exchange integrals involving the electrons with the same spin as electron 1. Briefly, ϵ_i consists of the kinetic energy [Eq. (2.8)] of an electron in orbital ϕ_i, plus the potential energy of interaction between that electron and the nucleus and all the other electrons.

Now consider the normalized Hartree–Fock spatial orbitals, namely, $\phi_{1s}, \phi_{2s}, \ldots,$ with energies $\epsilon_{1s}, \epsilon_{2s}, \ldots,$ respectively, occupied by $\nu_i \ (= 0, 1, \text{ or } 2)$ electrons. The potential energy of an electron with energy ϵ_i includes, on average, the repulsion between this electron and all the other electrons. The sum of the orbital energies $\sum_i \nu_i \epsilon_i$ thus counts each interelectronic repulsion twice. The Hartree–Fock energy of the atom E is therefore

$$E = \sum_i v_i \epsilon_i - V_{ee} \tag{2.13}$$

where V_{ee} is the interelectronic repulsion computed over the entire atom.

It is understood that all the integrals considered in this section are definite integrals over the full range of all the coordinates.

2.2 HARTREE–FOCK–ROOTHAAN ORBITALS

Following Roothaan's proposal, the Hartree–Fock orbitals are usually represented as linear combinations of a set of known basis functions χ_k^{lm}:

$$\phi_i = \sum_k c_{ki} \chi_k^{lm} \tag{2.14}$$

This representation permits analytic calculations, as opposed to fully numerical solutions [47,48] of the Hartree–Fock equation. Variational SCF methods using finite expansions [Eq. (2.14)] yield optimal analytic Hartree–Fock–Roothaan orbitals, and their corresponding eigenvalues, within the subspace spanned by the finite set of basis functions.

Commonly, one uses normalized Slater-type orbitals

$$\chi^{lm} = \frac{(2\zeta)^{n+1/2}}{[(2n)!]^{1/2}} r^{n-1} e^{-\zeta r} Y_l^m(\theta, \varphi) \tag{2.15}$$

where n is the principal quantum number and ζ is the orbital exponent. Alternatively, one can use large linear combinations of Gaussian functions

$$N r^l e^{-\zeta r^2} Y_l^m(\theta, \varphi) \tag{2.16}$$

which are particularly efficient in molecular calculations in that they require less computer time than do Slater integral evaluations. In atomic calculations, Slater functions are preferred because one-center integrals are no more difficult for Slater-type than for Gaussian-type orbitals and relatively few well-chosen Slater functions yield accurate results.

2.3 CONFIGURATION INTERACTION CALCULATIONS

To get true Hartree–Fock orbitals, an infinite set of basis functions should be included in the expansion (2.14). The question is: How can we improve our calculations using finite sets? In Hartree–Fock theory, the wavefunction of an n-electron atom that satisfies the antisymmetry requirement is a normalized Slater determinant D

$$
D = \frac{1}{\sqrt{n!}}
\begin{vmatrix}
\phi_1(1) & \phi_2(1) & \cdots & \phi_n(1) \\
\phi_1(2) & \phi_2(2) & \cdots & \phi_n(2) \\
\vdots & \vdots & & \vdots \\
\phi_1(n) & \phi_2(n) & \cdots & \phi_n(n)
\end{vmatrix}
\tag{2.17}
$$

where $\phi_1, \phi_2, \ldots, \phi_n$ are the orthonormal Hartree–Fock spin orbitals of that atom.

Configuration interaction (CI) is conceptually the simplest procedure for improving on the Hartree–Fock approximation. Consider the determinant formed from the n lowest-energy occupied spin orbitals; this determinant is $|\Psi_0\rangle$ and represents the appropriate SCF reference state. In addition, consider the determinants formed by promoting one electron from an orbital k to an orbital v that is unoccupied in $|\Psi_0\rangle$; these are the singly excited determinants $|\Psi_k^v\rangle$. Similarly, consider doubly excited ($k, l \rightarrow v, t$) determinants $|\Psi_{kl}^{vt}\rangle$ and so on up to n-tuply excited determinants. Then use these many-electron wavefunctions in an expansion describing the CI many-electron wavefunction $|\Phi_0\rangle$:

$$
|\Phi_0\rangle = c_0|\Psi_0\rangle + \sum_{k,v} c_k^v|\Psi_k^v\rangle + \sum_{k \geq l, v \geq t} c_{kl}^{vt}|\Psi_{kl}^{vt}\rangle + \cdots
\tag{2.18}
$$

Equation (2.18) is a linear variation function. (The summation indices prevent double-counting of excited configurations.) The expansion coefficients c_0, c_k^v, c_{kl}^{vt}, and so on are varied to minimize the variational integral. $|\Phi_0\rangle$ is a better approximation than $|\Psi_0\rangle$. In principle, if the basis were complete, CI would provide an exact solution. Here we use a truncated expansion retaining only determinants D' that differ from $|\Psi_0\rangle$ by at most two spin orbitals; this is a singly–doubly excited CI (SDCI).

The presence of excited determinants in $|\Phi_0\rangle$ introduces integrals of the type $\int D' \hat{A} D \, d\tau$, where $\hat{A}$ is an operator. Following the Condon–Slater rules, the n-electron integrals can be reduced to sums of one- and two-electron integrals [49]. Consider two determinants D and D', written as in (2.17), arranged so that as many as possible of their left-hand columns match. A one-electron operator $\hat{f}_i$ (viz., $-\frac{1}{2}\nabla_i^2$ or $-Z/r_i$) introduces the new integral

$$
\int \phi_n'(1)\hat{f}_1\phi_n(1)\,d\tau_1
\tag{2.19}
$$

when D and D' differ by one spin orbital, $\phi_n' \neq \phi_n$. No contribution arises when D and D' differ by two or more spin orbitals. The two-electron operator $|1/r_{12}|$ introduces the following new integrals

$$\sum_{j=1}^{n-1}\left[\int\int \phi_n'(1)\phi_j(2)\frac{1}{r_{12}}\phi_n(1)\phi_j(2)d\tau_1 d\tau_2 - \int\int \phi_n'(1)\phi_j(2)\frac{1}{r_{12}}\phi_j(1)\phi_n(2)d\tau_1 d\tau_2\right]$$

(2.20)

when D and D' differ by only one spin orbital, $\phi_n' \neq \phi_n$, or

$$\int\int \phi_n'(1)\phi_{n-1}'(2)\frac{1}{r_{12}}\phi_n(1)\phi_{n-1}(2)d\tau_1 d\tau_2 - \int\int \phi_n'(1)\phi_{n-1}'(2)\frac{1}{r_{12}}\phi_{n-1}(1)\phi_n(2)d\tau_1 d\tau_2$$

(2.21)

when D and D' differ by two spin orbitals: $\phi_n' \neq \phi_n$ and $\phi_{n-1}' \neq \phi_{n-1}$. No contribution arises when D and D' differ by three or more spin orbitals. The integrals (2.19)–(2.21) involve summation over the appropriate spin coordinates and integration over spatial coordinates.

Inspection of these integrals indicates that they are amenable to a space partitioning—like that involved in the atomic real-space core–valence separation described in Chapter 3—simply by selecting the appropriate limits of integration. Briefly, we can approach the study of core and valence regions with the help of CI wavefunctions.

Inclusion of the simply excited configurations is obviously needed for a proper description of one-electron properties such as charge densities and, hence, for properties that are sensitive to this quantity (dipole, quadrupole, etc. moments). The doubly excited configurations, on the other hand, are required for two reasons: (1) they allow us to indirectly introduce the single excitations by a coupling with the SCF reference state, as direct coupling is forbidden by the Brillouin theorem; and (2) they can lead to nonnegligible contributions to charge density corrections through a direct coupling with the reference state.

Examples

As one would expect, improvements are achieved for the energies, but, as usual in this type of work where the rate of convergence is rather slow, it is obvious that many configurations are required to approach accurate nonrelativistic solutions.

Regarding other properties, the influence of CI catches the eye. It is well known that the theoretical energy of ground-state molecules is only slightly improved by the inclusion of single excitations in CI calculations. In contrast, the important role of single excitations is revealed [50] by the erroneous dipole moment predicted for CO when only multiple excitations are retained; inclusion of single excitations restores the correct sign and order of magnitude of μ_{CO}. The main correction to the density arises from the interactions between single and double excitations and

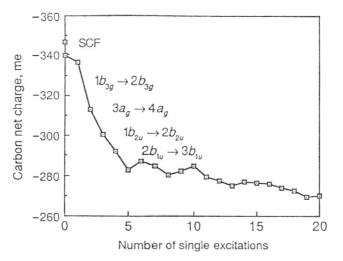

Figure 2.1. Contribution of single excitations, in decreasing order of their weight in the CI wavefunction, to the carbon net charge of ethylene.

between single excitations themselves, which reflects in essence the one-electron nature of the density operator.

The corrections are far more important for sp^2 than for sp^3 carbons. The charge analysis of ethylene offers a good example [51]. The results obtained by means of an optimized double-zeta 4-31G basis indicate that the SCF Mulliken carbon net charge of -346.4 me becomes -269.9 me in the SDCI calculation (see Fig. 2.1).

A core of two MOs was kept doubly occupied, namely, the $1a_g, 1b_{1u}$ orbitals corresponding to carbon K shells (g and u denote *gerade* and *ungerade*, respectively). The remaining 12 electrons were left available for fractional occupation of the 24 MOs.

In order to ensure the best possible wavefunction [Eq. (2.18)] while keeping the required computer space at an acceptable level, the configurations were selected by (1) retaining all singly excited symmetry- and spin-adapted configurations in the CI process and (2) selecting in order of preference those double excitations that mix more efficiently with single excitations and the SCF reference state. The latter rule introduces a new constraint, namely, that one-electron properties are thus expected to converge more rapidly than the correlation energy, subject only to the constraint embodied in the construction of c_k. Hence, a selection of configurations that may be appropriate in charge calculations does not necessarily represent the best choice in problems of correlation energies. Briefly, we approach convergence of the charge results well before achieving that of the energy. Convergence appears to be reached when all the single excitations are included in the configuration interaction and when the CI dimension reaches 400 determinants. (Similar remarks apply to acetylene as well. A selection of 17 single and 340 double excitations ensures convergence of the charge results. Three singly excited configurations of the $\sigma \to \sigma^*$ type account for $\sim 83\%$ of the net charge correction, while the $\pi \to \pi^*$ excitations are weak.)

TABLE 2.1. Contribution of Excited Configurations to Correlation Energy ΔE_{corr} and Mulliken Net Charges

Molecule	Level	c_0^2	ΔE_{corr} (au)	Δq_C (me)	q_C (me)
Methane	SCF	1.0	0	0.0	−523.2
	SDCI	>0.99	−0.07303	7.5	−515.7
Ethane	SCF	1.0	0	0.0	−382.7
	SDCI	>0.99	−0.05717	4.5	−378.2
Ethylene	SCF	1.0	0	0.0	−346.4
	DCI	0.93443	−0.10505	6.7	−339.7
	SDCI	0.93053	−0.10677	76.5	−269.9
	SC^a	—	−0.00172	69.8	—
Acetylene	SCF	1.0	0	0.0	−335.3
	DCI	0.92472	−0.13468	2.0	−333.3
	SDCI	0.92084	−0.13678	64.1	−271.2
	SC	—	−0.00209	62.1	—

^{a}SC = single contribution.

As revealed by Fig. 2.1, four configurations are responsible for $\sim$76% of the charge correction: $1b_{3g} \rightarrow 2b_{3g}$, $3a_g \rightarrow 4a_g$, $1b_{2u} \rightarrow 2b_{2u}$, and $2b_{1u} \rightarrow 3b_{1u}$. The improvement due to $\pi \rightarrow \pi^*$ excitations is minor. Additional information is offered in Table 2.1. The contribution of single excitations is important in unsaturated molecules. For alkanes, however, the SDCI corrections are minor, but not negligible.

The message is clear—the results are important. But we must await Chapter 5 before discussing them.

Calculations like those reported here—charge densities and dipole moments—are nowadays almost routine for relatively small molecules. The work involved to obtain the results displayed in Table 2.1 reveals that a good description of charges in highly symmetric molecules does not necessarily involve very large CI calculations. It is also clear, however, that atomic charges would be plagued by serious errors if CI corrections were left out. This being said, it is important to be aware that the charge results presented at this point should not be taken at face value; they still require additional corrections, as indicated in Chapters 5 and 8.

The SDCI calculations are somewhat more involved in calculations of atomic real-space core–valence partitioning models because of the two-center integrals (2.10) and (2.11) that require definite integration limits to cover the appropriate core and valence subspaces. Fortunately, these calculations are greatly aided by most efficient standard techniques.

CHAPTER 3

CORE AND VALENCE ELECTRONS

3.1 INTRODUCTION

The partitioning of ground-state atoms, ions, and molecules into core and valence regions reflects the old idea that the chemical properties are largely governed by the outer (or *valence*) electronic regions, that is, by what we shall call *valence electrons*. Although intuitively appealing, this partitioning is not cast in formal theory. The question as to whether a core–valence separation can be defined in a physically meaningful way is thus sensible. Undoubtedly, introduction of a suitable criterion is required to provide an acceptable operational definition.

A familiar way of handling this question is offered by the notion of *electronic shells*. By definition, an electronic shell collects all the electrons with the same principal quantum number. The K shell, for example, consists of $1s$ electrons, the L shell collects the $2s$ and $2p$ electrons, and so on. The valence shell thus consists of the last occupied electronic shell, while the *core* consists of all the inner shells. This segregation into electronic shells is justified by the well-known order of the successive ionization potentials of the atoms.

Now, in what is called *Hartree–Fock orbital space*—or simply *orbital space*—the total energy is partitioned from the outset into orbital energies, $\epsilon_i = \epsilon_{1s}, \epsilon_{2s}, \ldots$ Hence we can always consider a collection of electrons and deduce their total energy from the appropriate sum of their orbital energies, remembering, however, that one must also correct for the interelectronic repulsions which are doubly counted in any sum of Hartree–Fock eigenvalues. No special problem arises with

Atomic Charges, Bond Properties, and Molecular Energies, by Sándor Fliszár
Copyright © 2009 John Wiley & Sons, Inc.

core–valence separations in the orbital space, but it is still up to us to select the core (or valence) electrons as seems appropriate. It appears reasonable to use the order of orbital energies as a guideline and thus consider the $1s^2$ electrons as the core of the first-row elements or the $1s^2$, $2s^2$, $2p^6$ electrons for the second row. Briefly, we reencounter the familiar shell model.

Incidentally, let us mention that the essence of the "pseudopotential" methods [52] is to replace core electrons by an appropriate operator. The point is that the core–valence partitioning involved in these methods refers to the same orbital space as the corresponding all-electron calculations.

Now, what if we abandon the orbital-by-orbital electron partitioning in favor of a description based on the stationary ground-state electron density $\rho(\mathbf{r})$? Clearly, this will oblige us to redefine the core–valence separation. In sharp contrast with what was done in orbital space, we need a *partitioning in real space*. Let us begin with isolated atoms.

3.2   ATOMIC CORE AND VALENCE REGIONS

Here we consider an inner *spherical* core, centered at the nucleus, with radius r_b, and an outer valence region extending from r_b to infinity. The number of core electrons N^c is then

$$N^c = 4\pi \int_0^{r_b} r^2 \rho(r) dr \tag{3.1}$$

where $\rho(r)$ is the electron density at a distance r from the nucleus with charge Z. [We write $\rho(\mathbf{r}) = \rho(r)$ because of the spherical symmetry of the electronic density; see Ref. 53.] The definition of N^c now rests with the definition of the proper boundary r_b.

A number of suggestions were offered to that effect [54–61]. Let us briefly examine them.

Politzer [54–56] defines the *average ionization potential at the point* $\mathbf{r}$

$$\bar{I} = \sum_i \frac{\rho_i(\mathbf{r})|\epsilon_i|}{\rho(\mathbf{r})}$$

where $\rho_i(\mathbf{r})$ is the electron density of the orbital with energy ϵ_i and $\rho(\mathbf{r}) = \sum_i \rho_i(\mathbf{r})$ is the total electron density at the point $\mathbf{r}$. We can interpret $\bar{I}$ as the average energy required for the removal of one electron from the point $\mathbf{r}$ of an atom or a molecule. In ground-state atoms, $\bar{I}$ decreases in a piecewise manner along the coordinate r [56], and the regions between the inflexion points may be taken as electron shells. Indeed, the number of electrons contained in the sphere with radius r_b [Eq. (3.1)] are close to 2 for the first-row elements (e.g., 2.033 e for carbon and 2.030 e for neon) or close to 2 and 10 e for the larger atoms (e.g., 2.011 and 10.068 e for argon). These results are well substantiated for the atoms Li—Ca but less clear for atoms with d electrons, probably because of the interpenetration of subshells in the heavier atoms [55].

The "average local electrostatic potential" $V(r)/\rho(r)$, introduced by Politzer [57], led Sen and coworkers [58] to conjecture that the global maximum in $V(r)/\rho(r)$ defines the location of the core–valence separation in ground-state atoms. Using this criterion, one finds N^c values [Eq. (3.1)] of 2.065 and 2.112 e for carbon and neon, respectively, and 10.073 e for argon, which are reasonable estimates in light of what we know about the electronic shell structure. Politzer [57] also made the significant observation that $V(r)/\rho(r)$ has a maximum any time the radial distribution function $D(r) = 4\pi r^2 \rho(r)$ is found to have a minimum.

The minimum of this function, namely, $D(r)$, plays a major role in the Politzer–Parr approximation[1] [61] for the valence region energy of ground-state atoms

$$E^v = -\frac{3}{7}(Z - N^c)\int_{r_b}^{\infty} 4\pi r\rho(r)dr \qquad (3.2)$$

where the boundary surface separating the inner core and the outer valence regions is taken at $r_b = r_{min}$, that is, at the minimum of the radial distribution function. In Hartree–Fock calculations, minima of $D(r)$ occur approximately at the "right places" (in terms of the shell model), specifically, at $N^c \approx 2$ e for the first-row and at $N^c \approx 2$ and $N^c \approx 10$ e for the second-row elements. This result sheds light on the physical involvement of the electronic shell structure in a meaningful separation of an atom or ion into core and valence regions but should evidently not be taken too literally with Hartree–Fock wavefunctions.

The true merits of this approximation are discussed further below.

With reference to the minima of the radial distribution function $D(r)$, SCF analyses [61] using the near-Hartree–Fock wavefunctions of Clementi [64] indicate that the numbers of electrons found in the inner shell extending up to the minimum of $D(r)$ amount to $N^c = 2.054$ e (Be), 2.131 (C), 2.186 (O), 2.199 (F) and 2.205 electron (Ne). The results of Smith et al. [65] bearing on the boundaries in position space that enclose the exact number given by the Aufbau principle support the idea of "physical" shells compatible with that principle. The maxima of $D(r)$, on the other hand, also appear to be topological features indicative of shells, their positions correlate well with the shell radii from the Bohr–Schrödinger theory of an atom [66]. The critical points in $\nabla^2\rho(r)$, in contrast, although highly indicative of atomic shells in a qualitative sense, are not suitable for defining meaningful shell boundaries [67]. So, on the basis of these results, we shall keep in mind that the radial distribution function offers a vivid pictorial reference suggesting an involvement of the electronic

---

[1]In Thomas–Fermi theory, the ground-state energy of a neutral atom with nuclear charge Z is [62,63] $E = \frac{3}{7}Z^2\phi_0'$, with

$$\phi_0' = \left(\frac{\delta\phi(r)}{\delta r}\right)_{r=0} = \left(\frac{\delta(rV/Z)}{\delta r}\right)_{r=0}$$

where $V(r)$ is the total electrostatic potential at the distance r from the nucleus. Politzer and Parr applied this Thomas–Fermi formula to a hypothetical neutral atom containing $(Z - N^c)$ electrons in the field of an expanded effective nucleus of radius $r_b = r_{min}$ to get Eq. (3.2).

shell structure in the separation of core and valence regions in atoms, but also that one should definitely not attempt to carry this picture too far.

In comparison with experimental data, the inconvenience arising from fractional electron populations was compensated by interpolation [61]; the penultimate ionization potential (IP) was included with the $1s$ population in the valence region. Similarly, the $2s$ and $2p$ contributions were subtracted from the valence region by including the corresponding fractions of the larger IPs in the core. The same approximation was used for $3s$ and $3p$ electrons.

At long last, we are ready for an approach that is not based on the radial distribution function. With reference to the partitioning surface defined by r_b, the numbers of core electrons N^c and of valence electrons N^v are readily obtained by adequately integrating the electron density $\rho(\mathbf{r})$ between 0 and r_b or between r_b and ∞, respectively, using $dN = \rho\, d\tau = 4\pi r^2 \rho(r)\, dr$. Accordingly, the nuclear–electronic potential energies, $-Z \int [\rho(\mathbf{r})/r]\, d\tau$, are V_{ne}^c for the N^c core electrons and V_{ne}^v for the N^v valence electrons with kinetic energies T^c and T^v, respectively, obtained by appropriate selections of the integration limits. Now we come to the two-electron integrals. The interelectronic repulsion, V_{ee}, is split into three parts: (1) $V_{ee}^{cc} = $ electron–electron repulsion concerning only the charges of the core region; (2) V_{ee}^{vv}, the repulsion between valence electrons; and (3) V_{ee}^{cv}, the repulsion between the N^c core and the N^v valence electrons, with $V_{ee} = V_{ee}^{cc} + V_{ee}^{vv} + V_{ee}^{cv}$.

Now, the appropriate integration limits used in the calculation of $|1/r_{12}|$ interactions between electrons assigned to the core region τ^c and electrons associated with the valence space τ^v concern both Coulomb and exchange integrals. The latter are of particular interest, namely, the exchange that participates in the evaluation of V_{ee}^{cv}.

Consider the sum K^{cv}, which collects all the relevant exchange terms between the core electrons found in τ^c and the valence electrons found in τ^v. This sum can vanish; that is $K^{cv} = 0$ is possible. [The function $1s(1)2s(1)1s(2)2s(2)$, for example, can be positive or negative depending on whether r_1 and r_2 are on the same side or on opposite sides of the nodal surface. The final sign of this contribution thus depends on the locations of the boundary and nodal surfaces.]

K^{cv} depends on the radius selected for the inner core region and so does N^c, of course [68,69]. Thus we can represent K^{cv} as a function of N^c. Two examples are offered in Figs. 3.1 and 3.2. The contribution of the SDCI correction, although small, is clearly recognizable.

The results for neon are typical for first-row atoms, while those of argon are representative of second-row elements [69]; for the first-row elements, $K^c \approx 0$ for $N^c = 2$ e, whereas for the second-row atoms, $K^{cv} \approx 0$ for $N^c = 2$ and $N^c = 10$ e.[2]

It is certainly reassuring to find recognizable, nonarbitrary features suggesting a meaningful definition of boundary surfaces in real space. Nonzero exchange integrals between individual electrons are a well-known corollary of their indistinguishability. It thus seems natural to argue that a group of electrons should

---

[2]Self-consistent field results obtained with near-Hartree–Fock wavefunctions of Clementi and Roetti [70] indicate similar patterns for Ti, Cr, Fe, Ni, Zn, Ge, Se, and Kr, namely, "almost" vanishing K^{cv} integrals for $N^c = 2$ and $N^c = 10$ e [69]. A third point exists for krypton, for $N^c = 28$ e, where K^{cv} reaches a minimum [44].

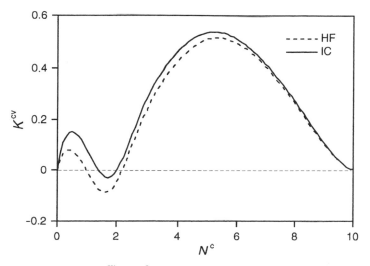

Figure 3.1. Neon. K^{cv} vs. N^c (au): $(13s\ 8p\ 2d\ 1f)$ SCF and SDCI results.

not be distinguished from another group of electrons if the total exchange between these groups is nonzero and that a vanishing K^{cv} should thus accompany a discrimination between core and valence electrons, as it is clearly illustrated in Figs. 3.1 and 3.2.[3]

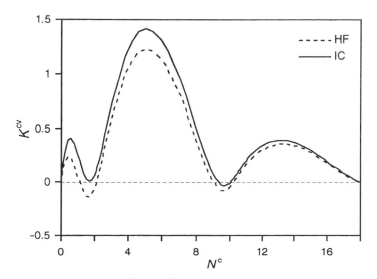

Figure 3.2. Argon. K^{cv} vs. N^c (au): 6-311G* SCF and SDCI results.

---

[3]From nickel onward, however, no boundary can be detected for $N^c = 28$ e—an observation that is consistent with a relatively significant degree of interpenetration, for third-row atoms, between the $3d$, $4s$, or $4p$ electrons and the $3s$ or $3p$ electrons, as shown by Politzer and Daiker [71].

TABLE 3.1. K^{cv} **Exchange Integrals (Atomic Units)**

Atom	N^c	Near-HF SCF	6-311G*	ANO	$(13s\ 8p\ 2d\ 1f)$
Li	2	−0.0027	−0.0024	−0.0021	−0.0017
Be	2	−0.0148	−0.0136	−0.0130	−0.0127
B	2	−0.0254	−0.0229	−0.0215	−0.0116
C	2	−0.0347	−0.0303	−0.0282	−0.0155
N	2	−0.0404	−0.0327	−0.0308	−0.0167
O	2	−0.0471	−0.0401	−0.0363	−0.0206
F	2	−0.0488	−0.0397	−0.0363	−0.0142
Ne	2	−0.0449	−0.0297	−0.0019	−0.0003
Na	10	0.0050	0.0064	0.0053	—
Mg	10	0.0118	0.0128	0.0124	—
Al	10	0.0158	0.0183	0.0204	—
Si	10	0.0171	0.0211	0.0204	—
P	10	0.0138	0.0182	0.0189	—
S	10	0.0019	0.0275	0.0077	—
Cl	10	−0.0186	−0.0111	−0.0098	—
Ar	10	−0.0450	−0.0075	−0.0358	—

Sources: Results taken from Ref. 69; the near-HF SCF Slater bases are from Ref. 70. The 6-311G* basis is that of Ref. 16. The contracted ANO (atomic natural orbital) bases [72] are indicated in [69], as well as the $(13s\ 8p\ 2d\ 1f)$ basis constructed with the help of van Duijneveldt's $(13s\ 8p)$ basis [18] augmented with d and f functions taken from Ref. 17.

Table 3.1 examines this point a little more in detail. For comparison, we offer the SCF results obtained with the near-Hartree–Fock wavefunctions of Clementi and Roetti [70] and the SDCI results obtained [69] with the 6-311G*, atomic natural orbital (ANO), and $(13s\ 8p\ 2d\ 1f)$ bases. It is clear that the K^{cv} integrals closely approach zero when $N^c = 2$ and $N^c = 10$ e. To appraise the significance of this result, we must compare it with the total exchange K^{total} calculated for the entire atom. K^{total} is small, of course, for lithium (0.0220 au) but increases rapidly with the size of the atom: 0.0587 (Be), 0.6396 (C), 1.2801 (O), and 2.3758 au (Ne) in the $(13s\ 8p\ 2d\ 1f)$ basis, with CI. Briefly, K^{total} is certainly sufficiently large to brand K^{cv} (for $N^c = 2$, viz., 10 e) as a negligible quantity, in comparison. Surely, this argument applies a fortiori also to the second-row elements where the total exchange integrals are still larger, up to ~ 7.2 au for argon.

It is difficult to ascertain conclusively whether small differences between K^{cv} and 0 stem from the incompleteness of our CI wavefunctions or whether they are (at least partly) genuine. So, while a small departure from 0 cannot be entirely ruled out for K^{cv} in situations where N^c is *exactly* 2 or 10 e, depending on what atom (or ion) we are talking about, it seems fair to claim that the criterion resting on vanishing core–valence exchange integrals clearly establishes the identity of the core and valence regions despite the possibly approximate nature of our numerical verifications. The following should also be considered.

Thus far we have dealt with the exchange between the N^c core and the N^v valence electrons, but did not consider the valence region energy E^v, the energy required for the removal of all the valence region electrons, thus leaving the 2-electron (or 10-electron) ion behind, with energy E^{ion}. From what we have learned here, only integer N^c and N^v populations need be considered, which makes comparisons with experimental data straightforward.

The appropriate formulas for E^v and E^{core} will be derived with these constraints in mind. But let us first clarify some points.

Computational Details

So far we have obtained a result regarding the boundary between core and valence regions in atoms, but little was said about the calculation of the relevant quantities. Useful information is given below.

Consider the partitioning of an atom (or ion) into two regions of space: a spherical inner region of radius r_b, centered at the nucleus, and an outer region, extending from r_b to infinity. The Hartree–Fock equation (2.2) is our starting point. Multiplication from the left by ϕ_i^*, integration from r_b to ∞, and summation over all occupied orbitals i leads to

$$\sum_i \int_{r_b}^{\infty} \nu_i \phi_i^* \hat{F} \phi_i \, d\tau = \sum_i \int_{r_b}^{\infty} \nu_i \phi_i^* \epsilon_i \phi_i \, d\tau \tag{3.3}$$

where ν_i is the occupation of the normalized orbital with eigenvalue ϵ_i. Until now, no constraint has been attached to the radius r_b defining the boundary surface separating the inner and outer regions. [The validity of (3.3) for all values of r_b depends on whether the ϕ_i values are true Hartree–Fock orbitals [73].]

Now we transform Eq. (3.3) into something more practical. Let us begin with its right-hand side. The integral

$$N_i^v = \nu_i \int_{r_b}^{\infty} \phi_i^* \phi_i \, d\tau \tag{3.4}$$

represents the number of electrons of orbital i found in the outer region (which we call for simplicity, but admittedly somewhat sketchily, the *valence region*). Equation (3.3) thus becomes

$$\sum_i \int_{r_b}^{\infty} \nu_i \phi_i^* \hat{F} \phi_i \, d\tau = \sum_i N_i^v \epsilon_i \tag{3.5}$$

Next, we proceed with the left-hand side of Eq. (3.5). Evaluation of the monoelectronic integrals occurring in it is straightforward. From Eq. (2.9), the nuclear–electronic potential energy of the $\sum_i N_i^v$ outer electrons in the field of the nuclear charge Z is

$$V_{ne}^{v} = -Z \sum_{i} v_i \int_{r_b}^{\infty} 4\pi r |\phi_i(r)|^2 \, dr \tag{3.6}$$

and from Eq. (2.8), their kinetic energy is,

$$T^{v} = \sum_{i} v_i \int_{r_b}^{\infty} \phi_i^* \hat{T} \phi_i \, d\tau. \tag{3.7}$$

Now we turn to the bielectronic integrals contained in the left-hand side of Eq. (3.5). They require a little attention. Consider the Coulomb operator $\hat{J}_j$ from Eq. (2.5). The integral $\int \cdots d\tau_2$ is over all space, but can be split into two contributions, namely, in short-hand notation, as

$$\int^{\tau_2^c} \cdots d\tau_2 + \int^{\tau_2^v} \cdots d\tau_2$$

where the first integral (from 0 to r_b) covers the inner (core) region τ_2^c, whereas the second one (from r_b to ∞) represents the outer (valence) space τ_2^v. Using Eq. (2.5), the integration between r_b and ∞ required by (3.5) thus gives the following Coulomb repulsion:

$$\int^{\tau_1^v} \phi_i^*(1)\hat{J}_j(1)\,\phi_i(1)\,d\tau_1 = \int\int^{\tau_1^v \tau_2^c} \cdots d\tau_2 \, d\tau_1 + \int\int^{\tau_1^v \tau_2^v} \cdots d\tau_2 \, d\tau_1 \tag{3.8}$$

These integrals describe a potential energy of interaction involving the part of electron 1 assigned to the valence space τ^v and a smeared-out electron with density $|\phi_j(2)|^2$. The two terms on the right-hand side of Eq. (3.8) differ from one another because the first integrals "sees" only the part of $|\phi_j(2)|^2$ found in the core space τ^c, whereas the second integral concerns only the part of $|\phi_j(2)|^2$ confined within the valence region τ^v.

At last, we can carry out the sum indicated in Eq. (3.5). This sum also includes a term arising from $\hat{J}_j(1)\phi_i(1)$:

$$\int^{\tau_2^v} \phi_j^*(2)\hat{J}_i(2)\phi_j(2)\,d\tau_2 = \int\int^{\tau_2^v \tau_1^c} \cdots d\tau_1 \, d\tau_2 + \int\int^{\tau_2^v \tau_1^v} \cdots d\tau_1 \, d\tau_2.$$

As a consequence, the summation over all occupied orbitals counts once again the 1−2 Coulomb interaction confined within the valence space, specifically, the $\int^{\tau_1^v} \int^{\tau_2^v} \cdots d\tau_2 \, d\tau_1$ term already found in (3.8), and adds the new integral

$$\int\int^{\tau_2^v \tau_1^c} \cdots d\tau_1 \, d\tau_2$$

describing the interaction between the part of electron 2 found in τ^v and the part of electron 1 found in the core region. Briefly, all interactions involving exclusively electron densities assigned to the outer space τ^v are counted twice in the sum (3.5), whereas the cross-interactions between electrons assigned to τ^v and those assigned to τ^c, represented by the appropriate $\int^{\tau_1^v} \int^{\tau_2^c} \cdots d\tau_2 \, d\tau_1 + \int^{\tau_2^v} \int^{\tau_1^c} \cdots d\tau_1 \, d\tau_2$ integrals, are counted only once. The same reasoning applies to the exchange operator (2.6) and to its integral $\int \cdots d\tau_1$, followed by summation over i. Down the line, taking now all the appropriate Coulomb and exchange integrals into account, it is seen that V_{ee}^{vv}, that is, the interelectronic repulsion involving only the electrons of the valence region, is counted twice in the left-hand side of Eq. (3.5) whereas V_{ee}^{cv}, namely, the repulsion between the core and the valence electrons, is counted only once.

The final result, including now V_{ne}^v and T^v, Eqs. (3.6) and (3.7), respectively, is therefore

$$\sum_i N_i^v \epsilon_i = T^v + V_{ne}^v + 2V_{ee}^{vv} + V_{ee}^{cv} \tag{3.9}$$

This equation [68,69,73] is a handy form of Eq. (3.3). All the terms can be evaluated by standard SCF and SDCI procedures, simply by paying attention to the appropriate limits of integration. (Full details are given in Refs. 68 and 69.) Evidently, all the quantities (except the ϵ_is) are functions of the r_b of our choice $[N_i^v = N_i^v(r_b), \ T^v = T^v(r_b),$ etc.]. A similar equation can be written for the core region by carrying out the integration of $\phi_i^* \hat{F} \phi_i$ between 0 and r_b, namely

$$\sum_i N_i^c \epsilon_i = T^c + V_{ne}^c + 2V_{ee}^{cc} + V_{ee}^{cv} \tag{3.10}$$

where T^c, V_{ne}^c, and V_{ee}^{cc} are, respectively, the kinetic, the nuclear–electronic, and the interelectronic energies of the $\sum_i N_i^c$ core electrons.

This concludes the presentation of the necessary formulas. Numerical examples are presented shortly.

3.3 THE VALENCE REGION ENERGY OF ATOMS

At last we can proceed with the derivation of the appropriate formulas for the core and valence region energies for both atoms and molecules. The atoms and their ions come first. This is another way of solving the question about the boundary separating the core and valence regions.

The valence region energy we are looking for, E^v, is *not* simply the sum of the kinetic and potential energies of the electrons in the outer (valence) region, as one would infer from their stationary densities. The valence energy described here accounts for any relaxation that accompanies an actual removal of the appropriate number of valence electrons. This energy should be a measure of the energy required to remove them. It carries the notion of integer numbers of electrons.

Relationship between Core and Valence Region Energies

Using Eq. (3.9), we can define a *virtual* valence region energy

$$\sum_i N_i^v \epsilon_i - V_{ee}^{vv} = T^v + V_{ne}^v + V_{ee}^{vv} + V_{ee}^{cv}$$

which is just the sum of the kinetic and potential energies associated with the valence region defined by r_b:

$$E_{virtual}^v = T^v + V_{ne}^v + V_{ee}^{vv} + V_{ee}^{cv} \qquad (3.11)$$

Similarly, we can define a *virtual* core energy:

$$E_{virtual}^c = T^c + V_{ne}^c + V_{ee}^{cc} + V_{ee}^{cv} \qquad (3.12)$$

Remembering that $E = T + V_{ne} + V_{ee}$ is the energy of the entire atom, it follows from Eqs. (3.11) and (3.12) that

$$E = E_{virtual}^v + E_{virtual}^c - V_{ee}^{cv} \qquad (3.13)$$

The valence energy (3.11) corresponds to a hypothetical ionization in which the valence electrons would be simply skimmed off as they are in the ground-state atom, with no relaxation of the core. This does not occur, of course, because electrons cannot be removed in fractional amounts. For that reason, $E_{virtual}^v$ is not an observable quantity.

An observable energy for a selected set of electrons follows in a straightforward manner from the sum of their orbital energies less the electron–electron interactions, which are counted twice in this sum. This procedure, reflecting the spirit of Koopman's theorem, is nothing new in Hartree–Fock theory.

So we go back and determine whether a valence energy other than (3.11) can be derived, complying with the requirement that it should represent a physical quantity. The problem at hand is best explained by an example. Suppose that there is some reason to assume that the core of carbon is adequately represented by a two-electron inner shell. In Hartree–Fock theory this core cannot consist of two pure $1s$ electrons; some $2s$ and $2p$ electron densities are found in this core, with some $1s$ density in the outer region. However, on ionization of the two $2s$ and the two $2p$ electrons, a C^{4+} ion is left behind with two pure $1s$ electrons. In other words, a relaxation is part of this process and is thus part of the description of the physical valence energy. Briefly, we have to look for a valence energy E^v by studying the general properties of an atom in which the core and valence regions are interdependent.

In this vein, we reexamine the problem with the understanding that in the event of an actual ionization, one region benefits at the expense of the other from the occurrence of relaxation. The valence and core energies modified by this relaxation are no longer those appearing in Eq. (3.13), but the modified valence region energy E^v and its core counterpart E^c are still tied by this general energy balance, that is

$$E = E^v + E^c - V_{ee}^{cv} \tag{3.14}$$

where E^v corresponds to the energy required for removal of the valence electrons, that is, to minus the sum of the relevant ionization potentials. When n valence electrons are removed from a neutral atom A, an ion with charge $+n$ is left behind. The same ion is formed on removal of $n + 1$ electrons from a negative ion A$^-$, or on removal of $n - 1$ electrons from A$^+$. Briefly, the valence energies of A$^-$, A, A$^+$, and so on are always expressed by reference to the same final ion. Inspection of Eq. (3.14) indicates that the ground-state energy of this ion is

$$E^{ion} = E^c - V_{ee}^{cv} \tag{3.15}$$

Energies E^v and E^c are the unknowns of our problem. Equation (3.14) is part of the solution.

Energy Formulas for Core and Valence Electrons

In the following we derive an expression for E^v without bothering about a physically valid core–valence separation, treating E^v as if it were a continuous function of r_b. The acceptable discrete solutions of E^v are selected afterward.

Consider first the electronic energy $E = \langle \Psi | \hat{H} | \Psi \rangle$ of a ground-state atom or ion with nuclear charge Z and apply the Hellmann–Feynman theorem [74] taking the nuclear charge Z as a parameter. This gives, in conventional notation, at constant electron density ρ

$$\left(\frac{\partial E}{\partial Z} \right)_\rho = \langle \Psi | \partial \hat{H} / \partial Z | \Psi \rangle = \left\langle \Psi \left| - \sum_i r_i^{-1} \right| \Psi \right\rangle$$

where the sum over i runs over all electrons. The well-known result is

$$\left(\frac{\partial E}{\partial Z} \right)_\rho = \frac{V_{ne}}{Z} \tag{3.16}$$

On the other hand, the virial theorem $(2E = V_{ne} + V_{ee})$ and the Hartree–Fock formula (2.13) combine to give

$$3E = V_{ne} + \sum_i v_i \epsilon_i \tag{3.17}$$

This result and Eq. (3.16) lead to

$$3E - Z\left(\frac{\partial E}{\partial Z}\right)_\rho = \sum_i v_i \epsilon_i \qquad (3.18)$$

We now define

$$\gamma = \frac{Z}{E}\left(\frac{\partial E}{\partial Z}\right)_\rho \qquad (3.19)$$

and rewrite Eq. (3.18) as follows:

$$(3 - \gamma)E = \sum_i v_i \epsilon_i \qquad (3.20)$$

Finally, combining (3.16) and (3.19), we get

$$E = \frac{1}{\gamma} V_{ne} \qquad (3.21)$$

Equations (3.20) and (3.21) represent an identity in Hartree–Fock theory. (The Hellmann–Feynman and virial theorems are satisfied by Hartree–Fock wavefunctions.) The particular interest offered by (3.21) lies in the fact that $\gamma = \frac{7}{3}$ appears to be the characteristic homogeneity of both Thomas–Fermi [62,75,76] and local density functional theory [77], in which case (3.20) gives the Ruedenberg approximation [78], $E = \frac{3}{2}\sum_i v_i \epsilon_i$, while (3.21) gives the Politzer formula [79], $E = \frac{3}{7} V_{ne}$. Equations (3.14), (3.20) and (3.21) are our working formulas.

We pick up things where we left them with Eq. (3.14) and write

$$\gamma E = \gamma^v E^v + \gamma^c (E^c - V_{ee}^{cv}) \qquad (3.22)$$

This equation is more general than what would follow from a simple multiplication by γ. It is noncommittal as to whether the atomic γ suits the individual core and valence parts; in writing (3.22), γ is taken as the average of γ^v (with a weight of E^v) and γ^c (weighted by $E^c - V_{ee}^{cv}$):

$$\gamma = \frac{\gamma^v E^v + \gamma^c (E^c - V_{ee}^{cv})}{E^v + E^c - V_{ee}^{cv}}$$

Next we consider $V_{ne} = V_{ne}^v + V_{ne}^c$, which we write as

$$V_{ne} = (V_{ne}^v + V_{ee}^{cv}) + (V_{ne}^c - V_{ee}^{cv})$$

The reason for this association of terms is physical. In a central-force problem, V_{ne}^v and V_{ee}^{cv} play roles that are similar in nature because V_{ne}^v measures the attraction of the valence electrons by the nucleus and V_{ee}^{cv} the accompanying repulsion by the core electrons. This formulation is directly linked to the model adopted here, that of a charged inner sphere of radius r_b surrounded by valence electrons. With

Eq. (3.21) in mind, we now write

$$\frac{1}{\gamma} V_{ne} = \frac{1}{\gamma^v} (V_{ne}^v + V_{ee}^{cv}) + \frac{1}{\gamma^c} (V_{ne}^c - V_{ee}^{cv}) \tag{3.23}$$

In this case $1/\gamma$ is the weighted average of $1/\gamma^v$ (with a weight of $V_{ne}^v + V_{ee}^{cv}$) and of $1/\gamma^c$ (with a weight of $V_{ne}^c - V_{ee}^{cv}$).

At this point we use Eqs. (3.22) and (3.23) and solve for γ^v. After some algebra one obtains $(\gamma^v)^2 + b\gamma^v + c = 0$ with $b = -[(V_{ne}^v + V_{ee}^{cv})/E^v + \gamma]$ and $b^2 - 4c = [(V_{ne}^v + V_{ee}^{cv})/E^v - \gamma]^2$. The trivial root is $\gamma^v = \gamma$. The other root gives

$$E^v = \frac{1}{\gamma^v} (V_{ne}^v + V_{ee}^{cv}) \tag{3.24}$$

So far we have exploited Eq. (3.21). We have one more step to go. Using (3.14) and (3.22), we write

$$(3 - \gamma)E = (3 - \gamma^v)E^v + (3 - \gamma^c)(E^c - V_{ee}^{cv})$$

and compare this expression with $\sum_i \nu_i \epsilon_i = \sum_i N_i^v \epsilon_i + \sum_i N_i^c \epsilon_i$. Equation (3.20) tells us that

$$\left[(3 - \gamma^v)E^v - \sum_i N_i^v \epsilon_i \right] + \left[(3 - \gamma^c)(E^c - V_{ee}^{cv}) - \sum_i N_i^c \epsilon_i \right] = 0$$

This equation achieves a core–valence separation. The terms in brackets are certainly individually zero at the limits $N^c = 0$ and $N^v = 0$, but this does not warrant that these terms are individually zero for other values of N^c, that is, that there is an N^c satisfying a meaningful core–valence separation. We shall tentatively proceed with

$$(3 - \gamma^v)E^v = \sum_i N_i^v \epsilon_i \tag{3.25}$$

and postpone momentarily the question about acceptable N^c values. Equation (3.25) defines E^v; it is the "valence counterpart" of Eq. (3.20).

Comparison with Eq. (3.24) yields the energy formula [68]

$$E^v = \frac{1}{3} \left(V_{ne}^v + V_{ee}^{cv} + \sum_i N_i^v \epsilon_i \right) \tag{3.26}$$

which is visibly the valence counterpart of (3.17). E^v expresses a valence region energy that takes the relaxation of the core into proper account. It corresponds to the appropriate sum of ionization potentials.

The final formula is thus obtained with the help of Eq. (3.9), namely

$$E^{\mathrm{v}} = \frac{1}{3}(T^{\mathrm{v}} + 2V^{\mathrm{v}}) \tag{3.27}$$

$$V^{\mathrm{v}} = V^{\mathrm{v}}_{\mathrm{ne}} + V^{\mathrm{vv}}_{\mathrm{ee}} + V^{\mathrm{cv}}_{\mathrm{ee}} \tag{3.28}$$

where V^{v} is the total potential energy of the $\sum_i N^{\mathrm{v}}_i$ electrons associated with the outer (valence) region. For the entire atom, that is, letting $N^{\mathrm{c}} = 0$, Eq. (3.27) reduces to $E = -T$ because $V/T = -2$ (virial theorem). While Eq. (3.9) is valid for any r_{b} of our choice, E^{v} is meaningful only for discrete values of N^{c}.

In closing, let us compare this physically observable E^{v} with $E^{\mathrm{v}}_{\mathrm{virtual}}$, Eq. (3.11), describing the "virtual" valence region of an unperturbed Hartree–Fock atom. We eliminate $V^{\mathrm{vv}}_{\mathrm{ee}}$ from (3.11) with the help of (3.9) and compare the result with (3.26). This gives

$$E^{\mathrm{v}}_{\mathrm{virtual}} = \frac{1}{2}(3E^{\mathrm{v}} + T^{\mathrm{v}}) \tag{3.29}$$

Undoubtedly, $E^{\mathrm{v}} \neq E^{\mathrm{v}}_{\mathrm{virtual}}$. Equation (3.29) indicates that the virial theorem, $-T^{\mathrm{v}} = E^{\mathrm{v}}$, and thus $-T^{\mathrm{v}} = E^{\mathrm{v}}_{\mathrm{virtual}}$, is not obeyed in the valence region.

Finally, one can also deduce the energy E^{ion} of the ion left behind on removal of the valence electrons (e.g., C^{4+} from C, C^{+}, or C^{-}) from the ground-state properties of its parent atom or ion. Using Eq. (3.10) and combining Eqs. (3.14), (3.15), and (3.22)–(3.25), we have

$$E^{\mathrm{ion}} = \frac{1}{3}\left(V^{\mathrm{c}}_{\mathrm{ne}} - V^{\mathrm{cv}}_{\mathrm{ee}} + \sum_i N^{\mathrm{c}}_i \epsilon_i\right) \tag{3.30}$$

$$E^{\mathrm{ion}} = \frac{1}{3}\left[T^{\mathrm{c}} + 2(V^{\mathrm{c}}_{\mathrm{ne}} + V^{\mathrm{cc}}_{\mathrm{ee}})\right] \tag{3.31}$$

Equations (3.26)–(3.29) still describe E^{v} as if it were a continuous function of r_{b}. There are restrictions, however, if E^{v} is meant to represent a physical quantity, namely, a valence energy that measures the energy actually required for the removal of integer numbers of outer electrons.

Examples: Meaningful Core–Valence Partitioning

The analysis presented earlier for the exchange integrals (Table 3.1 and Figs. 3.1 and 3.2) carries a strong conjecture regarding the uniqueness attached to the constraint $K^{\mathrm{cv}} = 0$, namely, concerning the validity of Eqs. (3.27) and (3.31) for appropriately selected core populations.

Typical examples, such as carbon, fluorine, and neon, nicely complete this argument. Table 3.2 lists their kinetic and potential valence region energy components for $N^{\mathrm{c}} = 1, 2, 3, \ldots$ electron and the corresponding E^{v} values given by Eq. (3.27); experimental data unmistakably pick $N^{\mathrm{c}} = 2$ e as the correct solution. Similar tests

TABLE 3.2. Selected Kinetic and Potential Valence Region Energies of C, F, and Ne (au)^a

Atom	N^c	V_{ee}^{vv}	V_{ee}^{cv}	V_{ne}^{v}	T^v	E^v Calculated	E^v Experimental^b
C	1	6.4902	4.5741	−36.1302	4.2833	−15.283	−19.849
	2	2.9648	5.7358	−18.0060	2.4073	−5.401	−5.440
	3	1.3611	4.6819	−10.4404	1.1533	−2.547	−3.070
F	1	26.0057	10.9725	−119.1931	20.9946	−47.812	−59.268
	2	16.8157	16.1570	−76.1107	13.4101	−24.289	−24.212
	3	10.9430	16.9484	−54.3146	8.7695	−14.692	−17.406
	4	6.8605	16.0133	−39.6055	4.6120	−9.617	−11.631
Ne	1	37.0049	13.7487	−163.0250	30.8793	−64.554	−78.991
	2	25.3052	20.8567	−109.1037	20.3326	−35.184	−35.045
	3	17.4010	22.8145	−80.7045	13.8571	−22.374	−26.258
	4	11.6797	22.5013	−61.1987	8.0261	−15.336	−18.641
	5	7.3925	20.6915	−45.6632	4.2586	−10.300	−12.837

^{a}SDCI results in the $(13s\ 8p\ 2d\ 1f)$ basis (from [69]).
^{b}Taken as the appropriate sum of ionization potentials [80], with a change in sign, using 1 au = 27.2106 eV.

select both $N^c = 2$ and $N^c = 10$ e for the second-row elements [68]. These constraints are now part of the definition of "effective" nuclear charges, $Z - N^c$.

Table 3.3 lists the relevant energy components of the first- and second-row atoms, for use in Eqs. (3.27) and (3.31). These are SDCI results obtained with the

TABLE 3.3. Kinetic and Potential Energies of First-Row Atoms ($N^c = 2$) and Second-Row Atoms ($N^c = 10$ e) (au)

Atom	V_{ee}^{cc}	V_{ee}^{vv}	V_{ee}^{cv}	V_{ne}^{c}	V_{ne}^{v}	T^c	T^v
Li	1.617	0.008	0.578	−16.239	−0.910	7.432	0.039
Be	2.333	0.303	1.752	−30.052	−3.649	14.372	0.283
B	3.072	1.183	3.436	−48.001	−8.968	23.649	0.990
C	3.848	2.965	5.736	−70.191	−18.006	35.415	2.407
N	4.654	5.955	8.653	−96.652	−31.728	49.734	4.819
O	5.484	10.466	12.089	−127.439	−50.646	66.674	8.343
F	6.337	16.816	16.157	−162.543	−76.111	86.255	13.410
Ne	7.212	25.305	20.857	−201.991	−109.104	108.521	20.333
Na	63.016	0.017	2.616	−386.486	−2.941	161.804	−0.002
Mg	72.414	0.276	6.824	−470.516	−8.352	199.500	0.093
Al	81.798	0.880	11.855	−562.875	−15.682	241.579	0.411
Si	91.237	2.004	18.275	−663.560	−25.996	288.028	0.987
P	100.785	3.771	25.990	−772.811	−39.572	339.054	1.883
S	110.425	6.389	34.798	−890.621	−56.465	394.672	3.093
Cl	120.233	9.943	44.917	−1016.857	−77.389	454.789	4.720
Ar	130.077	14.633	56.421	−1152.114	−102.854	519.993	6.832

SDCI results in the $(13s\ 8p\ 2d\ 1f)$ basis for Li—Ne and with the 6-311G* basis for the series Na—Ar.
Source: Ref. 69.

$(13s\ 8p\ 2d\ 1f)$ basis taking $N^c = 2$ e for the atoms Li—Ne and with Pople's 6-311G* basis for the second row, with $N^c = 10$ e. The latter basis was also used for the first-row atoms.

While the relatively modest 6-311G* set gives acceptable results for the first-row elements, things understandably deteriorate in the second row, particularly for the larger atoms. Our results are just fair for atoms larger than aluminum but nonetheless sufficiently clear to support the basic tenets underlying Eqs. (3.27) and (3.31) and our criterion defining core and valence regions. Of course, part of the problem is with the relatively modest size of the basis sets that were employed, but relativistic effects and size consistency certainly should be considered at this point. Relativistic corrections to the total energy are always negative for the ground-state configurations of atoms. In Datta's calculations [81] they amount to -0.01634 au for carbon, -0.05577 au for oxygen, and -0.14482 au for neon, to cite only a few examples. These corrections grow rapidly with the size of the atoms, for instance, -0.409 (Al), -0.771 (P), -1.024 (S), -1.339 (Cl), and -1.722 hartree for argon [82] and become numerically more important than possible improvements in CI calculations, a fact well worth remembering.

The results displayed in Tables 3.1–3.4 are self-explanatory: E^v and E^{ion} are meaningful only for discrete numbers N^c of electrons assigned to the core, namely, when the exchange integrals K^{cv} between N^c and N^v total (or at least closely approach) 0, that is, for $N^c = 2$ e or $N^c = 2$ and 10 e for the first- or second-row elements, respectively.

But the formulation (3.27) for E^v is not the most practical one for our intended applications to bond energy theory. A good approximation can be used instead.

TABLE 3.4. Calculated^a and Experimental Energies of Valence Electrons E^v and Two-Electron Ions E^{ion} (au)

Atom	6-311G*		$(13s\ 8p\ 2d\ 1f)$		Experimental^b	
	E^v	E^{ion}	E^v	E^{ion}	E^v	E^{ion}
Li	-0.201	-7.242	-0.202	-7.270	-0.198	-7.280
Be	-0.965	-13.662	-0.969	-13.688	-1.012	-13.657
B	-2.560	-22.047	-2.570	-22.070	-2.623	-22.035
C	-5.381	-32.404	-5.401	-32.424	-5.440	-32.416
N	-9.774	-44.735	-9.807	-44.754	-9.810	-44.802
O	-15.893	-59.056	-15.946	-59.079	-15.916	-59.194
F	-24.219	-75.358	-24.289	-75.386	-24.212	-75.595
Ne	-35.098	-93.642	-35.184	-93.679	-35.045	-94.006

^{a}Results obtained from Eqs. (3.27) and (3.31) for E^v and E^{ion}, respectively, using the input data given in Table 3.3 and similar data deduced with the help of Pople's 6-311G* basis set.
^{b}Taken as the appropriate sums of experimental ionization potentials [80], with a change in sign, using the conversion factor 1 au = 27.2106 eV.
Source: Ref. 69.

Alternate Energy Formula

Consider Eq. (3.24) and examine the physical nature of its V_{ee}^{cv} part. This potential energy describes the repulsion between an outer electron cloud (the valence electrons) and the N^c core electrons. A considerable simplification can be achieved with the help of Gauss' theorem, which offers a simple solution: Gauss' theorem tells us that if the charge of a subshell is rigorously spherically symmetric about a nucleus, the effect felt by a point charge "outside" such a subshell would be as though all the "inner" charge were concentrated at the nucleus to form a positively charged core that acts as a point charge.

Hence, with the $\int_{r_b}^{\infty} [\rho(r)/r]\,d\mathbf{r}$ integrals the same in the evaluation of V_{ee}^{cv} and V_{ne}^{v}, and accounting for the fact that the nuclear charge Z has been replaced by N^c, it follows that

$$V_{ee}^{cv} = -\frac{N^c}{Z} V_{ne}^{v} \tag{3.32}$$

It is now clear that

$$V_{ne}^{v} + V_{ee}^{cv} = -(Z - N^c) \int_{r_b}^{\infty} \frac{\rho(r)}{r}\,d\mathbf{r} \tag{3.33}$$

represents the nuclear–electronic potential energy of the outer electrons in the field of an expanded "effective nucleus" $(Z - N^c)$. This is an approximation, of course, because it does not consider the spin of the core electrons. Now we know that the exchange part K^{cv} of the core–valence interactions is certainly very small for the r_b that defines the boundary between core and valence regions. These results, which are in general agreement with those of Politzer and Daiker [71], suggest that (3.33), although approximate, is not bad at all. Note that because of Eq. (3.32), we can also write

$$V_{ne}^{v} + V_{ee}^{cv} = \frac{Z - N^c}{Z} V_{ne}^{v} \tag{3.34}$$

In defense of Eqs. (3.33) and (3.34), one can add that monoelectronic properties are better reproduced in Hartree–Fock theory than are bielectronic properties.

The simplification introduced with the use of Gauss' theorem is most valuable for the physical picture it conveys, that of a valence electron cloud in the field of a nucleus partially screened by its core electrons. Using it in Eq. (3.24), we get

$$E^v = -\frac{1}{\gamma^v}(Z - N^c) \int_{r_b}^{\infty} \frac{\rho(r)}{r}\,d\mathbf{r} \tag{3.35}$$

Remembering that, given spherical symmetry, it is $d\mathbf{r} = 4\pi r^2\,dr$, we see that Eq. (3.35) is our counterpart of the Politzer–Parr Thomas–Fermi-like formula (3.2) describing the valence region of atoms. Here we must stress that *Eq. (3.35) represents a valence energy in which relaxation effects are included.* It was derived from Eq. (3.24) and cannot be mistaken for $E_{\text{virtual}}^{\text{v}}$—an important aspect which was not recognizable in earlier derivations of energy formulas like (3.35) in the spirit of Thomas–Fermi theory.

This derivation foreshadows a similar one for molecules.

3.4 SUMMARY

The formula for the valence region energy in *real space*, namely, $E^{\text{v}} = \frac{1}{3}(T^{\text{v}} + 2V^{\text{v}})$, sharply differs from that applicable in the *orbital space* where the valence energy is just the simple sum of the pertinent electronic kinetic and potential energies. The relevant kinetic and potential energies are evidently not the same in the orbital space and in real space. They proceed from integrations over the full coordinate space in the former, thus contrasting with integrals spanning appropriate portions of the total space for use in $E^{\text{v}} = \frac{1}{3}(T^{\text{v}} + 2V^{\text{v}})$.

In orbital space, the core–valence separation of electrons is made with reference to some property (orbital energy or principal quantum number), but in real space this segregation is made solely with reference to the admissible number N^{c} of core electrons. The uniqueness of this partitioning in real space (which carries over in orbital space where the valence energies are the same) goes back to E^{v}, which accepts only discrete solutions, namely, with $N^{\text{c}} = 2$ e for the first-row elements or $N^{\text{c}} = 2$ and 10 e for the second row, that is, whenever the exchange integrals between the core and the valence region electrons are down to zero. For any other N^{c}, large discrepancies are found between the calculated E^{v} values and the corresponding ionization potential sums; the agreement between E^{v} and the sum of the ionization potentials provides strong support for the contention that physically meaningful cores are defined in real space by taking $N^{\text{c}} = 2$ e for the first-row atoms and $N^{\text{c}} = 2$ or 10 e for the larger elements.

Finally, this valence region energy can be expressed, with the help of Gauss' theorem, by an equivalent, considerably simplified approximation that features only the nuclear–electronic potential energy of the valence electrons in the field of an expanded nucleus partially screened by its core electrons.

CHAPTER 4

THE VALENCE REGION OF MOLECULES

4.1 MODEL

This chapter is about molecules. A *molecule* is a collection of nuclei $Z_k, Z_l, \ldots$ at rest (in the Born–Oppenheimer approximation) in a sea of fast-moving electrons. The nuclei can be identified and thus provide convenient reference marks. Each nucleus found in the molecule, say, Z_k, can be viewed as the nucleus of a "giant atom" extending over the entire molecule: Z_k is surrounded by charges, just as in an ordinary atom, with the difference that motionless positive point charges are now part of its environment. Of course, the electronic content is still described by its stationary density.

This abridged description hints at the strategy adopted in our work but requires clarification. When an atom becomes part of a molecule, it is clear that both its nucleus *and* its electrons are incorporated. So, when we say that Z_k is in a molecule, we refer more precisely to what we shall call "atom k in the molecule," meaning that (1) Z_k has entered the molecule together with its electrons—those of the isolated atom k—and (2) that the incorporated atom k differs from the isolated atom k. This vision of an atom in a molecule does not introduce any physical constraint. The molecule is still taken as a whole, as described above. No spatial partitioning of the molecule is considered, carrying the picture of subspaces defined by boundaries enclosing one nucleus and a share of the electronic charge entirely distributed among the subspaces. The only charge partitioning contemplated here concerns a separation into core and valence regions along the lines described for the isolated atoms.

Atomic Charges, Bond Properties, and Molecular Energies, by Sándor Fliszár
Copyright © 2009 John Wiley & Sons, Inc.

The argument is developed in three steps:

1. The bare nucleus, Z_k, of atom k in the molecule is considered in the field of all the electrons and of all the other nuclei found in the molecule. This introduces the notion of binding.

2. Z_k is considered together with its core electrons N_k^c. This amounts to a core–valence partitioning of "atom k in the molecule"; its valence region now consists of all the electrons except N_k^c and of all the other nuclei. At this stage it is not taken into account that the other nuclei—those embedded in the valence region of atom k—may possess core electrons of their own.

3. The final step considers all the nuclei with their own core electrons, meaning that all the remaining electrons shall be regarded as the valence electrons of the molecule. The idea is to find an expression for the energy of a molecule featuring the role of its electronic valence region.

The valence region energy of ground-state atoms or ions—that is, with a change in sign, the energy required to remove the valence electrons—is given by Eq. (3.27), and for the ion left behind after removal of the valence electronic charge, we write Eq. (3.31).

Here we wish to show that *the same real-space formulas apply to molecules* as well, but V^v has to be redefined because it must now incorporate the internuclear repulsion energy V_{nn} and also accommodate more than one single core. Concerning E^{ion} and the terms appearing in Eq. (3.31), however, they need not be redefined. With E_k^{ion} for the energy of the kth ionic core (say, H^+, C^{4+}, N^{5+}, O^{6+}) and E for the molecule—all energies referring to ground states—the valence energy E^v under consideration satisfies the important constraint

$$E = E^v + \sum_k E_k^{ion} \tag{4.1}$$

So we begin with E^v, in real space. Then, for the purpose of setting our approach in perspective, we discuss its relation to the familiar methods in orbital space. Finally, we examine the relative merits of the two models. Motivation is drawn from the fact that real-space philosophy largely governs atom-by-atom and bond-by-bond descriptions of molecules, while current core–valence separation schemes for molecules are rooted in orbital space theory, with no provision whatsoever for real-space applications.

That should be remedied.

4.2   THE CORE–VALENCE SEPARATION IN REAL SPACE

In writing Eq. (4.1), we assume that all particles, the molecule and the ions, are at rest and that the molecule is in its equilibrium geometry. The total energy of the latter, $E = \langle \Psi | \hat{H} | \Psi \rangle$, is expressed in the Born–Oppenheimer approximation. The

Hamiltonian $\hat{H}$ is for an n-electron molecule, in atomic units

$$\hat{H} = -\frac{1}{2}\sum_i \nabla_i^2 - \sum_k \sum_i \frac{Z_k}{r_{ik}} + \sum_j \sum_{i>j} \frac{1}{r_{ij}} + \sum_k \sum_{l>k} \frac{Z_k Z_l}{R_{kl}}$$

where i and j refer to electrons and k and l refer to nuclei. The first term is the operator for the kinetic energy of the electrons, where ∇_i^2 is the familiar Laplacian operator

$$\nabla_i^2 \equiv \frac{\partial^2}{\partial x_i^2} + \frac{\partial^2}{\partial y_i^2} + \frac{\partial^2}{\partial z_i^2}$$

The second term represents the attraction between the nuclei and the electrons, where r_{ik} is the distance between electron i and nucleus k. The third term represents the repulsion between electrons i and j at a distance r_{ij}. The last term represents the repulsion between the nuclei, where R_{kl} is the distance between nuclei with charges Z_k and Z_l.

The derivative $(\partial E/\partial Z)_\rho$ at constant electron density ρ with respect to the nuclear charge of one of the nuclei is obtained with the help of the Hellmann–Feynman theorem [74]. So we get the potential at the center k, namely, $V_k/Z_k = (\partial E/\partial Z_k)_\rho$, and the corresponding potential energy

$$V_k = Z_k \left(\frac{\partial E}{\partial Z_k} \right)_\rho$$

$$= -Z_k \int \frac{\rho(\mathbf{r})}{|\mathbf{r} - \mathbf{R}_k|} \, d\mathbf{r} + Z_k \sum_{l \neq k} \frac{Z_k}{R_{kl}} \tag{4.2}$$

where $\rho(\mathbf{r})$ is the electron density in the volume element $d\mathbf{r}$ at the point $\mathbf{r}$ and $\mathbf{R}_k$ defines the position of nucleus Z_k. Summation over all centers k gives

$$\sum_k V_k = V_{\text{ne}} + 2V_{\text{nn}} \tag{4.3}$$

$$V_{\text{nn}} = \sum_k \sum_{l>k} \frac{Z_k Z_l}{R_{kl}} \tag{4.4}$$

where V_{nn} is the familiar internuclear repulsion and

$$V_{\text{ne}} = -\sum_k Z_k \int \frac{\rho(\mathbf{r})}{|\mathbf{r} - \mathbf{R}_k|} \, d\mathbf{r} \tag{4.5}$$

measures the nuclear–electronic attraction energy.

The molecule is taken in its equilibrium geometry. So we apply the molecular virial theorem, $2E = V_{\text{ne}} + V_{\text{ee}} + V_{\text{nn}}$, where V_{ee} is the interelectronic repulsion.

Using Eqs. (4.2) and (4.3), we get

$$2E - \sum_k Z_k \left(\frac{\partial E}{\partial Z_k}\right)_\rho = V_{ee} - V_{nn} \tag{4.6}$$

On the other hand, in Hartree–Fock theory the total energy is $E = \sum_i v_i \epsilon_i - (V_{ee} - V_{nn})$, where v_i is the occupation (0, 1, or 2) of orbital i with eigenvalue ϵ_i. Combining this result with (4.6) to get rid of its $(V_{ee} - V_{nn})$ part, we write

$$(3 - \gamma)E = \sum_i v_i \epsilon_i \tag{4.7}$$

where γ has been defined as

$$\gamma \equiv \frac{1}{E} \sum_k Z_k \left(\frac{\partial E}{\partial Z_k}\right)_\rho \tag{4.8}$$

Equations (4.2), (4.3), and (4.8) indicate that

$$E = \frac{1}{\gamma}(V_{ne} + 2V_{nn}) \tag{4.9}$$

for a molecule. For isolated atoms, this equation reduces to Eq. (3.21).

Finally, for an atom k embedded in a ground-state molecule, the corresponding expression is

$$E_k = \frac{1}{\gamma_k} V_k \tag{4.10}$$

subject to the constraint that the average of the $(1/\gamma_k)$ values, weighted by V_k, must restore the $1/\gamma$ of Eq. (4.8):

$$\frac{1}{\gamma} = \frac{\sum_k (1/\gamma_k) V_k}{\sum_k V_k} \tag{4.11}$$

This definition of γ_k leads to

$$E = \sum_k E_k \tag{4.12}$$

and thus also to the result (4.9) indicated above.

Binding

What is E_k? It is tempting to call it the energy of an atom in a molecule. We do so for the sake of simplicity, but shall not forget that the subscript k associates E_k with an atom that is identified solely by its nucleus Z_k. There is one E_k for each nucleus. These E_k values have the desirable property that their sum is the total energy of the molecule. Therein lies their usefulness, unabated by the fact that the individual E_k terms are not amenable to direct measurements [9].

The energy E_k differs, in principle, from the energy of the isolated ground-state atom k, which is E_k^{atom}. The difference

$$\Delta E_k = E_k^{atom} - E_k \tag{4.13}$$

contains part of the molecular binding energy. A good insight into the meaning of ΔE_k is offered by the atomization energy ΔE_a^*, defined as

$$\Delta E_a^* = \text{energy of all the isolated ground-state atoms}$$
$$\text{minus the ground-state energy of the molecule}$$
$$= \sum_k E_k^{atom} - E^{mol} \tag{4.14}$$

where $E^{mol} \equiv E$. It follows immediately from Eqs. (4.12)–(4.14) that

$$\sum_k \Delta E_k = \Delta E_a^* \tag{4.15}$$

Equation (4.15) is important, it offers an atom-by-atom partitioning of the molecular binding energy. The nice thing about (4.15) is that it does not imply any spatial partitioning of the molecule. Equation (4.13) is instrumental in the theory of bond energies. ΔE_a^* is convenient for comparisons with experimental results.

The Real-Space Core–Valence Partitioning

Now we can proceed with this topic.

The potential energy to be used in Eq. (4.10) is that given in Eq. (4.2). Now we rewrite Eq. (4.2) but introduce two modifications:

1. First we separate the nuclear–electronic potential energy contributed by the core electrons associated with Z_k, which we call $V_{ne,k}^c$, from that due to all the electronic charge found outside the core region of atom k.
2. Next we consider that the core electrons associated with Z_k do interact with the charges found outside that core. On one hand, they repel these "external" electrons and thus reduce their effective attraction by nucleus Z_k. This attraction by Z_k and the concurrent repulsion by N_k^c play similar roles, one interaction opposing the other, and are considered jointly. On the other hand, the core electrons N_k^c attract the nuclei $Z_l \ldots$ and thus counteract the repulsion between Z_k and the other nuclei. These repulsions and counteracting attractions also belong together. In short, the core electrons screen not only the attraction between Z_k and the outer electrons but also the internuclear repulsion involving Z_k.

The total screening imputable to N_k^c is written

$$V_k^{cv} = \text{interaction energy between } N_k^c \text{ core electrons}$$
$$\text{and electronic and nuclear charges found outside}$$
$$k\text{th core containing } Z_k \text{ and } N_k^c$$

The form of Eq. (4.2) that reflects this model is

$$V_k = \left[-Z_k \int_{r_{b,k}}^{\infty} \frac{\rho(\mathbf{r})}{|\mathbf{r} - \mathbf{R}_k|} \, d\mathbf{r} + Z_k \sum_{l \neq k} \frac{Z_k}{R_{kl}} + V_k^{cv} \right] + (V_{ne,k}^c - V_k^{cv}) \tag{4.16}$$

So we go back to Eq. (4.10), use Eq. (4.16), and write E_k as

$$E_k = \frac{1}{\gamma_k^v} \left[-Z_k \int_{r_{b,k}}^{\infty} \frac{\rho(\mathbf{r})}{|\mathbf{r} - \mathbf{R}_k|} \, d\mathbf{r} + Z_k \sum_{l \neq k} \frac{Z_l}{R_{kl}} + V_k^{cv} \right] + \frac{1}{\gamma_k^c} (V_{ne,k}^c - V_k^{cv}) \tag{4.17}$$

where the $(1/\gamma_k)$ parameter of Eq. (4.10) is treated as the average of $1/\gamma_k^v$ (weighted by the term in brackets) and of $1/\gamma_k^c$ (with a weight of $V_{ne,k}^c - V_k^{cv}$). This description of E_k is noncommital as to whether the γ_k of Eq. (4.10) suits the individual core and valence parts.

The first part of the right-hand side (RHS) of Eq. (4.17) represents the valence region energy of atom k embedded in the molecule, and the second term is the energy $E_k^{ion} = (V_{ne,k}^c - V_k^{cv})/\gamma_k^c$ of the ionic core k. The total energy $E = \sum_k E_k$ is thus

$$E = \frac{1}{\gamma^v} \sum_k \left[-Z_k \int_{r_{b,k}}^{\infty} \frac{\rho(\mathbf{r})}{|\mathbf{r} - \mathbf{R}_k|} \, d\mathbf{r} + Z_k \sum_{l \neq k} \frac{Z_l}{R_{kl}} + V_k^{cv} \right] + \sum_k E_k^{ion} \tag{4.18}$$

where $1/\gamma^v$ is the appropriate average of the individual $(1/\gamma_k^v)$ values, weighted by the terms in brackets in Eq. (4.18). The first RHS term of Eq. (4.18) describes the valence region energy of a molecule: E^v. Equation (4.18) is a form of Eq. (4.1) and highlights the role of potential energies. It is instrumental in the derivation of the final energy formulas that lead to practical bond energy formulas. But let us first complete the derivation of the appropriate formula for E^v.

4.3   FORMULA FOR THE VALENCE REGION ENERGY

At this point we have all the required tools for deriving E^v.

Up to now, potential energies were at the center of our arguments. Little attention was paid to the electronic kinetic energy. This situation arose from our application of the Hellmann–Feynman theorem with intent to stress the role of the potential energies—a role made explicit in Eq. (4.18)—while the kinetic energy component was seemingly neglected. In fact, it is somehow hidden in γ^v. We shall now calculate γ^v and thus reintroduce explicitly the appropriate valence electronic kinetic energy T^v into the formula describing E^v.

We begin with Eq. (4.1) and write, with Eqs. (4.16)–(4.18) in mind, the following equation:

$$\gamma E = \gamma^v E^v + \gamma^c \sum_k E_k^{ion} \tag{4.19}$$

[The RHS of Eq. (4.19) is, from Eqs. (4.17) and (4.18) and comparison with Eq. (4.3), equal to $V_{ne} + 2V_{nn}$, which is γE.] Now use Eqs. (4.1) and (4.19) and write

$$(3 - \gamma)E = (3 - \gamma^v)E^v + (3 - \gamma^c)\sum_k E_k^{ion} \tag{4.20}$$

The $(3 - \gamma)E$ term is well known [see Eq. (4.7)]. In the latter, ν_i is the occupation of the orbital whose energy is ϵ_i. Here we use $\nu_i = N_i^c + N_i^v$, so that $(3 - \gamma)E = \sum_i N_i^v \epsilon_i + \sum_i N_i^c \epsilon_i$. Consequently, we deduce from Eq. (4.20) that

$$\left[(3 - \gamma^v)E^v - \sum_i N_i^v \epsilon_i\right] + \left[(3 - \gamma^c)\sum_k E_k^{ion} - \sum_i N_i^c \epsilon_i\right] = 0 \tag{4.21}$$

The latter equation achieves a separation of the core and valence contributions. The terms in brackets are individually zero at the limits $N^c = 0$ and $N^v = 0$. Here we postulate that physically meaningful core populations exist that allow such a core–valence separation and proceed with

$$(3 - \gamma^v)E^v = \sum_i N_i^v \epsilon_i \tag{4.22}$$

for the valence region. Now we compare this expression with the E^v appearing in Eq. (4.18), eliminate γ^v, and obtain

$$E^v = \frac{1}{3}\left[\sum_k\left(-Z_k\int_{r_{b,k}}^{\infty}\frac{\rho(\mathbf{r})}{|\mathbf{r} - \mathbf{R}_k|}\,d\mathbf{r} + Z_k\sum_{l \neq k}\frac{Z_l}{R_{kl}} + V_k^{cv}\right) + \sum_i N_i^v \epsilon_i\right] \tag{4.23}$$

This formula can be simplified. The first term on its RHS is $V_{ne}^v = V_{ne} - \sum_k V_{ne,k}^c$, specifically, the total nuclear–electronic potential energy of the molecule [Eq. (4.5)] stripped of all the individual core nuclear–electronic interactions $V_{ne,k}^c$. Next, we decompose $\sum_k V_k^{cv}$:

$$\sum_k V_k^{cv} = V_{ee}^{cv} + 2V_{ee}^{inter} + V_{ne}^{inter} \tag{4.24}$$

where V_{ee}^{cv} is the repulsion between core and valence electrons. V_{ee}^{inter} is the core–other core interelectronic repulsion and V_{ne}^{inter} is the core–other nucleus attraction.

So we get from (4.23) that

$$E^{\mathrm{v}} = \frac{1}{3}\left(V_{\mathrm{ne}}^{\mathrm{v}} + V_{\mathrm{ne}}^{\mathrm{inter}} + V_{\mathrm{ee}}^{\mathrm{cv}} + 2V_{\mathrm{ee}}^{\mathrm{inter}} + \sum_i N_i^{\mathrm{v}}\epsilon_i \right) \qquad (4.25)$$

where E^{v} is almost in its final form. The last step concerns $\sum_i N_i^{\mathrm{v}}\epsilon_i$.

We start with the Hartree–Fock formula (2.2) as indicated earlier and proceed from Eq. (3.3) to Eq. (3.5), which is now written

$$\sum_i \int^{\mathrm{val}} \nu_i \phi_i^* \hat{F}\phi_i\, d\tau = \sum_i N_i^{\mathrm{v}}\epsilon_i \qquad (4.26)$$

The left-hand side (LHS) of Eq. (4.26) will tell us what to use in Eq. (4.25) instead of $\sum_i N_i^{\mathrm{v}}\epsilon_i$. Concerning the one-electron terms, the integrals carried out solely over the valence space yield the nuclear–electronic potential energy $V_{\mathrm{ne}}^{\mathrm{v}} - V_{\mathrm{ne}}^{\mathrm{inter}}$ of the N^{v} valence electrons plus the kinetic energy T^{v} of the same. For the two-electron integrals, the LHS of (4.26) collects all the pertinent Coulomb and exchange terms between the valence electrons and those assigned to the cores in $V_{\mathrm{ee}}^{\mathrm{cv}}$, as well as the repulsions involving exclusively valence electrons, $V_{\mathrm{ee}}^{\mathrm{vv}}$. A double-counting of the latter occurs in the summation over all i values:

$$\sum_i N_i^{\mathrm{v}}\epsilon_i = T^{\mathrm{v}} + V_{\mathrm{ne}}^{\mathrm{v}} - V_{\mathrm{ne}}^{\mathrm{inter}} + V_{\mathrm{ee}}^{\mathrm{cv}} + 2V_{\mathrm{ee}}^{\mathrm{vv}} \qquad (4.27)$$

Finally, we use this expression in Eq. (4.25) and obtain

$$E^{\mathrm{v}} = \frac{1}{3}[T^{\mathrm{v}} + 2(V_{\mathrm{ne}}^{\mathrm{v}} + V_{\mathrm{ee}}^{\mathrm{vv}} + V_{\mathrm{ee}}^{\mathrm{cv}} + V_{\mathrm{ee}}^{\mathrm{inter}} + V_{\mathrm{nn}})] \qquad (4.28)$$

The sum $V_{\mathrm{ee}}^{\mathrm{v}} = V_{\mathrm{ee}}^{\mathrm{vv}} + V_{\mathrm{ee}}^{\mathrm{cv}} + V_{\mathrm{ee}}^{\mathrm{inter}}$ represents the total interelectronic repulsion stripped of all core contributions: $V_{\mathrm{ee}} - \sum_k V_{\mathrm{ee},k}^{\mathrm{cc}}$. Thus we obtain the final result from Eq. (4.28), [83]:

$$E^{\mathrm{v}} = \frac{1}{3}(T^{\mathrm{v}} + 2V^{\mathrm{v}})$$

$$V^{\mathrm{v}} = V_{\mathrm{ne}}^{\mathrm{v}} + V_{\mathrm{ee}}^{\mathrm{v}} + V_{\mathrm{nn}} \qquad (4.29)$$

The valence region of molecules and of the isolated ground-state atoms or ions are described by the same formula, Eq. (3.27). For isolated atoms, of course, we use $V_{\mathrm{ee}}^{\mathrm{inter}} = 0$, $V_{\mathrm{ne}}^{\mathrm{inter}} = 0$, and $V_{\mathrm{nn}} = 0$. The description presented here for the molecules is a generalization of that offered earlier for the atoms and contains the latter as a special case.

Our formula for E^{ion} [Eq. (3.31)] proceeds from the same general approach, using $(3 - \gamma^c) \sum_k E_k^{\text{ion}} = \sum_i N_i^c \epsilon_i$, from Eq. (4.21), along the lines outlined for the calculation of E^{v}.

4.4 INTERFACE WITH THE ORBITAL MODEL

In the valence molecular calculations, the total molecular energy can be decomposed as follows:

$$E = E_{\text{valence}} + E_{\text{core}} + \sum_{k>l} \sum \frac{Z_k Z_l}{|\mathbf{R}_k - \mathbf{R}_l|} \qquad (4.30)$$

Here, E_{core} is the energy contribution from the core orbitals. If the latter are classified according to the nuclear center on which they are located (e.g., on nucleus k), the set of core orbitals belonging to this center is $\{\phi_c, c \in k\}$. Moreover, if these core orbitals are assumed to be nonoverlapping, the core energy may be partitioned into two terms [84]:

$$E_{\text{core}} = E_{\text{core}}^{(1)} + E_{\text{core}}^{(2)} \qquad (4.31)$$

The one-center term

$$E_{\text{core}}^{(1)} = \sum_k \left[\sum_{c \in k} 2\langle \phi_c | -\frac{1}{2}\nabla^2 - \frac{Z_k}{|\mathbf{r} - \mathbf{R}_k|} | \phi_c \rangle + \sum_{c \in k} \sum_{c' \in k} (2J_{cc'} - K_{cc'}) \right] \qquad (4.32)$$

is the sum of the Hartree–Fock core energies associated with each center k, so that we can identify $E_{\text{core}}^{(1)} = \sum_k E_k^{\text{ion}}$. The two-center term

$$E_{\text{core}}^{(2)} = 2 \sum_{k \neq l} \sum_{c \in l} \langle \phi_c | -\frac{Z_k}{|\mathbf{r} - \mathbf{R}_k|} | \phi_c \rangle + \sum_{k \neq l} \sum_{c \in k} \sum_{c' \in l} (2J_{cc'} - K_{cc'}) \qquad (4.33)$$

collects the core–other nucleus attraction terms as well as the core–other core repulsions. $E_{\text{core}}^{(2)}$ vanishes when the cores are infinitely separated. At this point, a comparison of Eq. (4.1) with Eq. (4.30) shows that

$$E^{\text{v}} = E_{\text{valence}} + E_{\text{core}}^{(2)} + \sum_{k>l} \sum \frac{Z_k Z_l}{|\mathbf{R}_k - \mathbf{R}_l|} \qquad (4.34)$$

Now we go along with an argument offered by Truhlar et al. [84]. In the evaluation of Eq. (4.33), and consistent with the nonoverlapping core orbitals assumption, we can neglect the core–other core exchange interactions. Because the core charge densities $\rho_k(\mathbf{r}) = 2 \sum_{c \in k} \phi_c^*(\mathbf{r})\phi_c(\mathbf{r})$ are spherically symmetric about their nuclear centers and the cores are assumed to be nonoverlapping, one obtains the approximation [84]

$$E_{core}^{(2)} + \sum\sum_{k>l} \frac{Z_k Z_l}{|\mathbf{R}_k - \mathbf{R}_l|} \cong \sum\sum_{k>l} \frac{(Z_k - N_k^c)(Z_l - N_l^c)}{|\mathbf{R}_k - \mathbf{R}_l|} \qquad (4.35)$$

and thus

$$E^v \cong E_{valence} + \sum\sum_{k>l} \frac{(Z_k - N_k^c)(Z_l - N_l^c)}{|\mathbf{R}_k - \mathbf{R}_l|} \qquad (4.36)$$

In this approximation, the net effect of the core interaction energy stands for the shielding of the nuclear charges in the internuclear repulsion.

Let us briefly state where we stand. $E_{valence}$ denotes occupied valence orbitals only and is simply represented by the straight sum of their pertinent kinetic and potential energies computed over the entire coordinate space. E^v, in contrast, described by $\frac{1}{3}(T^v + 2V^v)$, represents all occupied orbitals but integrated only over specified (core and valence) regions of space. The relationship between the two, Eq. (4.36), is surprisingly simple considering the basic differences between the two models.

Numerical Examples

Table 4.1 lists selected CI results showing that addition or withdrawal of one electron to or from an electroneutral atom has little effect on the energy components of its electronic cores. It thus seems a reasonable approximation to consider the neutral atom values as reference for the forthcoming calculations. Table 4.2 lists the pertinent GTO[$5s\,3p$] results used in conjunction with molecular calculations carried out with

TABLE 4.1. Kinetic and Potential Energies^a of Core Electrons of Selected Atoms and Ions, A, A$^+$, and A$^-$, for Use in Eq. (3.31) (au)

Atom (Ion)	T^c	V_{ee}^{cc}	V_{ne}^c	E^{ion} Calculated	E^{ion} Experimental^b
C$^+$	35.4638	3.8308	−70.3472	−32.523	−32.416
C	35.3696	3.8513	−70.1708	−32.423	
C$^-$	35.2478	3.8510	−70.0409	−32.377	
N$^+$	49.8039	4.6395	−96.8377	−44.864	−44.802
N	49.6647	4.6572	−96.6275	−44.759	
N$^-$	49.5478	4.6568	−96.5146	−44.723	
O$^+$	66.7254	5.44741	−127.6279	−59.194	−59.194
O	66.5554	5.4863	−127.4102	−59.098	
O$^-$	66.4056	5.4861	−127.2674	−59.052	

^{a}SDCI results obtained [69] with the ANO (atomic natural orbital) [$7s\,6p\,3d$] basis given in Ref. 86.
^{b}Taken as minus the appropriate sum of ionization potentials [80].

TABLE 4.2. Core Energies of Selected Atoms (au)[a]

Atom	T^c	V_{ee}^{cc}	V_{ne}^c	E^{ion}	$E_{core}^{(1)}$
C	35.3844	3.9253	−70.1985	−32.3873	−32.3602
N	49.6513	4.7353	−96.6647	−44.7358	−44.7339
O	66.6077	5.5665	−127.4352	−59.0432	−59.1071

[a]SCF results obtained with Dunning's GTO[$5s\,3p|3s$] basis [85]. The $E_{core}^{(1)}$ energies are numerical HF evaluations of Eq. (4.32), from Ref. 87.

Dunning's GTO[$5s\,3p|3s$] basis [85] to get E^{ion}. The corresponding $E_{core}^{(1)}$ energies, deduced from Eq. (4.32), are also indicated.

Numerical Hartree–Fock calculations of $E_{core}^{(1)}$ [87], on the other hand, convincingly show that *our results in real space are the same as those of the orbital space model* [Eq. (4.32)] and that we are thus justified to write

$$E_k^{ion} = E_{core(k)}^{(1)} \qquad (4.37)$$

for each individual center k. Pertinent $E_{core}^{(1)}$ values from Eq. (4.32) [87] are indicated in Table 4.2. These Hartree–Fock results are indeed close to those given by Eq. (3.31). Additional Hartree–Fock results are [87] -75.4797 au (F) and -444.7455 au (Cl), compared with -75.386 au and -446.153 au, respectively, from SDCI calculations of E^{ion} using Eq. (3.31) [69], and experimental values of -75.595 au and -446.356 au, respectively, from the appropriate sums of ionization potentials.

The identification of Eq. (4.37) is important because it suffices to establish the link between E^v and $E_{valence}$ [Eq. (4.34)], deduced from Eqs. (4.1), (4.30), and (4.31). Moreover, if these numerical results are now taken as a validation—without explanation, of course—of our formula for E^{ion} [Eq. (3.31)], it follows from Eq. (4.1) that E^v must have the form given in Eq. (3.27), with $E = \frac{1}{3}(T + 2V)$ for the total energy, which reverts to the standard formula $E = T + V$ with the use of the virial theorem. (Note, however, that the virial theorem is not satisfied for the core and valence subsystems taken individually, i.e., $E^v \neq -T^v$ and $E^c \neq -T^c$.)

The valence region kinetic energy T^v is readily obtained by subtracting all the appropriate core kinetic energies from the calculated total kinetic energy. Similarly, one obtains V_{ne}^v from the total nuclear–electronic potential energy from which we subtract all the pertinent core V_{ne}^c terms. Finally, we deduce V_{ee}^v from the total interelectronic repulsion energy, from which we subtract the pertinent V_{ee}^{cc} terms. The internuclear repulsion V_{nn}, of course, is that obtained by carrying out the usual optimizations of the total molecular energy. The final results are displayed in Table 4.3.

These are SCF results obtained with the GTO($9s\,5p|6s$) → [$5s\,3p|3s$] basis using Dunning's exponents [85] and optimum contraction vectors [85]. The "error" (ΔE) represents the difference between our calculation using Eqs. (4.1) and (3.27) and the genuine SCF result; it is positive whenever $|V/T| > 2$ and negative when $|V/T| < 2$.

TABLE 4.3. Application of Eqs. (3.27) and (4.1) for Selected Molecules (au)

Molecule	T^v	V_{ne}^v	V_{ee}^v	V_{nn}	E^v	ΔE^a
C_2	4.6357	−65.3522	32.0608	15.0712	−10.6015	−0.0140
CH_4	4.8390	−49.8395	22.1775	13.5245	−7.8120	−0.0120
C_2H_4	7.3098	−108.0649	50.8928	33.6233	−13.2626	−0.0182
C_2H_6	8.4502	−127.7946	59.6461	42.2663	−14.4380	−0.0030
C_3H_8	12.0952	−228.8817	108.6595	82.5597	−21.0765	−0.0047
i-C_4H_{10}	15.7683	−353.4116	169.3705	134.5693	−27.7250	−0.0155
i-C_4H_8	14.5551	−320.7158	154.1553	119.4861	−26.5311	−0.0059
C_6H_6	18.3307	−522.1176	255.1082	203.3607	−36.3221	0.0047
N_2	9.3840	−107.7574	51.1787	22.8683	−19.3456	0.0654
NH_3	6.4401	−58.4697	26.4138	11.7185	−11.4116	0.0280
N_2H_2	10.7276	−129.8096	61.2135	32.4678	−20.5070	−0.0257
N_2H_4	11.6968	−149.7488	70.3847	41.0728	−21.6286	0.0505
HCN	7.7719	−98.2720	46.9096	23.9108	−15.7104	0.0131
CH_2N_2	13.1203	−204.4055	97.6572	61.2808	−25.9382	−0.0049
NH_2CN	13.3304	−202.4348	96.8968	59.7849	−26.0586	−0.0498
CH_3CN	11.3545	−186.4146	89.3394	57.8880	−22.3400	0.0380
O_2	16.2673	−155.1460	72.2121	27.6771	−31.4155	0.0098
H_2O	9.4319	−71.5823	32.2030	9.1950	−16.9789	−0.0087
CO	10.6469	−112.1393	52.8103	22.1348	−21.2472	0.0194
CO_2	19.0035	−234.4352	110.8640	58.4157	−37.1025	−0.0134
H_2CO	11.8005	−132.6016	62.1461	30.9690	−22.3908	0.0145
CH_3OH	13.0380	−153.8342	71.7371	40.1893	−23.5925	−0.0034
$(CH_3)_2O$	16.6620	−261.2893	123.8250	83.8112	−30.2148	−0.0028
$(C_2H_5)_2O$	24.0549	−515.2101	247.6725	190.2076	−43.5350	−0.0361
N_2O	17.3578	−226.8735	107.8693	57.8640	−34.9742	0.1106

[a]ΔE is taken as the difference $E^v + \sum_k E_k^{ion} - E_{SCF}^{mol}$.

4.5   APPROXIMATION FOR THE VALENCE ENERGY

The evaluation of E^v presents no difficulty, but there is another, simplified, and most useful form that catches our attention.

In Thomas–Fermi theory, adoption of the simple central field model for neutral molecules at equilibrium leads to simple energy relations—well supported by SCF calculations [12]—such as $E = \frac{3}{7}(V_{ne} + 2V_{nn})$. Evidently, nothing of the like applies to $E_{valence}$, but we may well inquire how things are with E^v.

The key is in the treatment of core–other core and core–other nucleus interactions. Simple approximations were presented in that matter to get Eq. (4.35). Assuming Gauss' theorem—the V_{ee}^{cv} potential is just as though all the core electronic charge were lumped at the nuclear position—the same arguments are now invoked for V_k^{cv}, approximated as follows:

$$V_k^{cv} = N_k^c \int_{r_{b,k}}^{\infty} \frac{\rho(\mathbf{r})}{|\mathbf{r} - \mathbf{R}_k|} \, d\mathbf{r} - N_k^c \sum_{l \neq k} \frac{Z_l}{R_{kl}} \qquad (4.38)$$

Direct calculations [89] made for $1s$ electrons confirm the validity of Eq. (4.38).

The first RHS term of Eq. (4.38) describes the repulsion between N_k^c, located at the point $\mathbf{R}_k$, and all the outer electrons. The second term describes the attraction between N_k^c and all the nuclei other than Z_k. Equation (4.18) now becomes

$$E^v = \frac{1}{\gamma^v} \sum_k \left[-Z_k^{\text{eff}} \int_{r_{b,k}}^{\infty} \frac{\rho(\mathbf{r})}{|\mathbf{r} - \mathbf{R}_k|} \, d\mathbf{r} + Z_k^{\text{eff}} \sum_{l \neq k} \frac{Z_l}{R_{kl}} \right] \tag{4.39}$$

where $Z_k - N_k^c$ defines the *effective* nuclear charge Z_k^{eff}. The integral appearing in Eq. (4.39) runs over the entire space outside the boundary $r_{b,k}$, hence also over regions containing the core electrons of the other nuclei. Consider instead a "truncated" integral $\int^{\text{val}} \cdots d\mathbf{r}$ that avoids systematically the core electrons of all atoms, including N_k^c, as

$$\int_{r_{b,k}}^{\infty} \frac{\rho(\mathbf{r})}{|\mathbf{r} - \mathbf{R}_k|} \, d\mathbf{r} = \int^{\text{val}} \frac{\rho(\mathbf{r})}{|\mathbf{r} - \mathbf{R}_k|} \, d\mathbf{r} + \sum_{l \neq k} \frac{N_l^c}{R_{kl}}$$

and rewrite Eq. (4.39) as follows:

$$E^v = \frac{1}{\gamma^v} \sum_k \left[-Z_k^{\text{eff}} \int^{\text{val}} \frac{\rho(\mathbf{r})}{|\mathbf{r} - \mathbf{R}_k|} \, d\mathbf{r} + Z_k^{\text{eff}} \sum_{l \neq k} \frac{Z_l^{\text{eff}}}{R_{kl}} \right] \tag{4.40}$$

Then note that

$$\sum_k Z_k^{\text{eff}} \int^{\text{val}} \frac{\rho(\mathbf{r})}{|\mathbf{r} - \mathbf{R}_k|} \, d\mathbf{r} = V_{\text{ne}}^{\text{eff}} \tag{4.41}$$

appropriately describes the total *effective* nuclear–electronic potential energy of the molecule, that is, the potential energy of its valence electrons—and only those—in the field of the effective nuclear charges Z_k^{eff}, Z_l^{eff}, and so forth. The summation of the nuclear repulsion terms in Eq. (4.40) gives $2V_{\text{nn}}^{\text{eff}}$, where $V_{\text{nn}}^{\text{eff}} = \sum_k \sum_{l>k} Z_k^{\text{eff}} Z_l^{\text{eff}} / R_{kl}$ is the total repulsion between Z_k^{eff}, Z_l^{eff}, ... The final result is thus

$$E^v = \frac{1}{\gamma^v} \left(V_{\text{ne}}^{\text{eff}} + 2V_{\text{nn}}^{\text{eff}} \right) \tag{4.42}$$

This equation describes chemical binding in the simplest possible way, in terms of *effective potentials* at the nuclei. Therein lies the importance of this approximation, as shown in the next section. It lays the foundation of the bond energy theory, which will be developed shortly.

Numerical Examples

Let us now turn to Eq. (4.42). The nuclear and electronic potentials at the nuclei and the appropriate V_{ne}^c values give access to V_k^{cv} through the approximation expressed in Eq. (4.38). So, as the required internuclear repulsion energies are known, we can evaluate $V_{ne}^{eff} + 2V_{nn}^{eff}$ for use in Eq. (4.42). Finally, comparison with the corresponding E^v values obtained from Eq. (3.27) enables the evaluation of the γ^v parameters of Eq. (4.42). [The calculation of $\sum_k V_k^{cv}$ is practical if we proceed as indicated in Eq. (4.18). Alternatively, we can skip this step and use Eq. (4.39), which is a form of Eq. (4.42).] Selected results are given in Table 4.4.

The $1/\gamma^v$ parameters introduced with Eq. (4.18) represent averages of individual "atomic" $1/\gamma^v$ parameters weighted by *effective* electronic and nuclear potential energies, namely, the terms in brackets shown in Eq. (4.39) or (4.40)—a situation similar to that prevailing in all-electron applications of $E = (1/\gamma)(V_{ne} + 2V_{nn})$. In the latter case, $1/\gamma$ ($\approx 3/7$) is the average of "atomic" $1/\gamma_k$ terms weighted by the *total* potential energies V_k, Eq. (4.2), with $1/\gamma_k = 0.500$ (H), 0.429 (C), 0.426 (N), and 0.422 (O) [90].

TABLE 4.4. Calculation of $\sum_k V^{cv}$ and $V_{ne}^{eff} + 2V_{nn}^{eff}$, to Obtain γ^v from (4.42) (au)^a

Molecule	$\sum_k V_k^{cv}$	$V_{ne}^{eff} + 2V_{nn}^{eff}$	γ^v
C_2	11.7366	-23.4732	2.2141
CH_4	6.1128	-16.6772	2.1349
C_2H_6	12.1784	-31.0835	2.1529
C_3H_8	18.2606	-45.5017	2.1589
i-C_4H_{10}	24.3511	-59.9218	2.1613
C_2H_4	12.1455	-28.6721	2.1623
i-C_4H_8	24.2924	-57.4512	2.1654
C_2H_2	12.0906	-26.2128	2.1823
C_6H_6	36.2814	-79.1149	2.1781
N_2	17.7202	-44.3006	2.2900
NH_3	9.1173	-25.9155	2.2710
N_2H_4	18.1233	-49.4800	2.2877
N_2H_2	17.9593	-46.9107	2.2875
HCN	14.9936	-35.4568	2.2569
CH_2N_2	23.6326	-58.2112	2.2442
NH_2CN	23.9638	-58.9012	2.2603
CH_3CN	20.9833	-49.6553	2.2227
O_2	24.9479	-23.4732	2.3824
H_2O	12.8120	-40.3821	2.3784
CO	18.4241	-49.4456	2.3272
CO_2	30.7941	-86.8096	2.3397
H_2CO	18.6000	-52.0636	2.3252
CH_3OH	18.7830	-54.6725	2.3174
H_2CCO	24.4384	-63.6231	2.2834
$(CH_3)_2O$	24.7477	-68.9192	2.2810
$(C_2H_5)_2O$	36.9388	-97.8560	2.2478
N_2O	29.9487	-81.1968	2.3216

^{a}SCF results obtained with the $(9s\,5p|6s) \rightarrow [5s\,3p|3s]$ basis [85,88].

Here, in our valence region applications, we find "atomic" $1/\gamma^v$ terms of 0.500 (H), 0.455 (C), 0.436 (N), and 0.421 (O). On the basis of the present $[5s\,3p|3s]$ SCF results alone, it is difficult to judge conclusively how accurate our approximations are, namely, that proposed for V_k^{cv}, Eq. (4.38), but backcalculations using these $1/\gamma^v$ parameters and the appropriate SCF potential energies shown in Eq. (4.39) reproduce the results given by Eq. (3.27) typically within $\pm0.15\%$ or better, which seems acceptable. The γ^v values of Table 4.4 show that while hydrogen tends to lower them because of the nonnegligible contribution of $1/\gamma_k^v = 0.500$, the observed values are generally of the order anticipated from Thomas–Fermi theory. This situation parallels that encountered in all-electron calculations of γ [90].

We shall now proceed with Eq. (4.42) and show the importance of this approximation.

4.6 PERTURBATION OF THE VALENCE REGION

The formula for the ground-state energy of a molecule follows from Eq. (4.42):

$$E^{mol} = \frac{1}{\gamma^v}\left(V_{ne}^{eff} + 2V_{nn}^{eff}\right) + \sum_k E_k^{ion} \tag{4.43}$$

Consider also the valence region energy $E_k^{v,atom}$ of the isolated ground-state atoms $k,l,$ ... and rearrange Eq. (4.43) as follows:

$$\sum_k E_k^{ion} + \sum_k E_k^{v,atom} - E^{mol} = \sum_k E_k^{v,atom} - \frac{1}{\gamma^v}\left(V_{ne}^{eff} + 2V_{nn}^{eff}\right) \tag{4.44}$$

The total energy of an isolated ground-state atom k is $E_k^{atom} = E_k^{ion} + E_k^{v,atom}$. Equation (4.14) thus tells us that the LHS of (4.44) is ΔE_a^*, the energy of atomization of the molecule. So we write

$$\Delta E_a^* = \sum_k E_k^{v,atom} - \frac{1}{\gamma^v}\left(V_{ne}^{eff} + 2V_{nn}^{eff}\right) \tag{4.45}$$

where ΔE_a^* is the central quantity involved in any comparison with thermochemical data, such as enthalpies of formation, enthalpies of atomization, and the like, and gives the total ground-state energy of the molecule with the help of Eq. (4.14). ΔE_a^* is what we want to calculate.

Suppose that we solve Eq. (4.45) by making some assumptions—namely, regarding "model" electron densities, $\rho^\circ(\mathbf{r})$ and internuclear distances R_{kl}° taken from a suitable reference system—and write

$$\Delta E_a^{*\circ} = \sum_k E_k^{v,atom} - \frac{1}{\gamma^v}\left(V_{ne}^{eff} + 2V_{nn}^{eff}\right)^\circ \tag{4.46}$$

The superscript "∘" identifies quantities computed for the model. Hence, combining now Eqs. (4.45) and (4.46), we deduce the important formula

$$\Delta E_a^* = \Delta E_a^{*\circ} - \frac{1}{\gamma^v} \Delta \left(V_{ne}^{eff} + 2V_{nn}^{eff} \right) \tag{4.47}$$

The perturbation $\Delta(V_{ne}^{eff} + 2V_{nn}^{eff})$ describes the replacement of model densities and internuclear distances by the values that are appropriate for the molecule under scrutiny. Similarly, appropriate "reference" atomic energies must be used in the atomic-like formula (4.15) to get $\Delta E_a^{*\circ}$. Ingeniously selected references require small corrections. Nature helps a lot in that matter by keeping the changes of $\rho(\mathbf{r})$ as small as possible.

The bond energy theory is rooted in Eq. (4.47).

4.7 SUMMARY

The partitioning of the electronic charge of ground-state molecules into core and valence parts unfolds as a straightforward extension of the methods applied to ground-state atoms. In *real space* the valence region energy E^v satisfies the condition $E = E^v + \sum_k E_k^{ion}$, where E is the total ground-state energy of the molecule (or atom) under scrutiny and E_k^{ion} is that of the ion k (such as H^+, C^{4+}, N^{5+}, etc.) left behind on removal of the entire valence region electronic charge of atom k. E^v thus includes relaxation, that is, whatever energy changes occur when the "valence electrons" are added to the ions k *forced into adopting the equilibrium geometry of an incipient ground-state molecule.* So it justifies, albeit in an unconventional way, its designation "valence energy."

For ion k it is shown that $E_k^{ion} = \frac{1}{3}[T^c + 2(V_{ne}^c + V_{ee}^{cc})]$, where T^c is the kinetic energy of its N^c core electrons, V_{ne}^c their nuclear–electronic potential energy, and V_{ee}^{cc} their interelectronic repulsion. In *orbital-space* SCF theory, on the other hand, E^{ion} is identified with the appropriate sum of its Hartree–Fock core orbital energies, that is, $E^{ion} = E_{core}^{(1)}$, a result supported by independent direct numerical HF calculations.

As a direct consequence, the familiar orbital-space valence energy $E_{valence}$ and the present real-space E^v energy are related to one another in a straightforward manner, namely, by the approximation $E^v = E_{valence} + V_{nn}^{eff}$, which highlights the role of the repulsions between nuclei partially screened by their core electrons. $E_{valence}$ is the usual straight sum of the kinetic and potential energies of the pertinent occupied "valence orbitals" computed over the entire coordinate space, whereas E^v denotes all occupied orbitals, but integrated only over specified (core and valence) regions of real space. The relationship between E^v and $E_{valence}$ offers nothing beyond what seems obvious on simple physical grounds; the novelty is in the formulation, $E^v = \frac{1}{3}(T^v + 2V^v)$, of the valence energy in the present real-space partitioning scheme, where T^v and V^v are, respectively, the relevant kinetic and potential energies.

The present description means that E^v accounts for chemical binding, which makes it particularly attractive in real-space applications to molecules. Indeed, any atom-by-atom or bond-by-bond partitioning of a molecule, such as Bader's "atom-in-the-molecule" model, is by its very nature treated in real space. In this particular context, we benefit from the fact that the present real-space core–valence partitioning correctly reflects chemical binding, as indicated by the atomic-like formula $\sum_k \Delta E_k = \Delta E_a^*$ for the atomization energy ΔE_a^*, where ΔE_k is the energy difference between a free atom and a bonded atom k. The atoms-in-a-molecule concept is compulsively needed in chemistry, but ambiguous and subject to personal choice, as seen here.

One cannot but adhere to the conclusion of Parr et al. [91]:

> The term "noumenon" (in the sense of Kantian philosophy) is deadly correct for describing atoms-in-a-molecule as a conceptual construct ultimately unknowable by observation or unique definition, but conceivable by reason. Chemical science is built upon the atom, and the atom in a molecule is a vital, central concept, yet forever elusive: there are multiple ways to partition molecules into atoms that are consistent with various observed chemical trends and experimental data.

Our strategy was dictated by the specificity of the bond energy problem to be solved and the considerable simplification thus achieved; it is just one facet of the reality. We owe the model to the work of Politzer and Parr [9]. In a nutshell:

> The Politzer–Parr partitioning of molecular energies in terms of "atomic-like contributions" results in an exact formula for the nonrelativistic ground-state energy of a molecule as a sum of atomic terms that emphasizes the dependence of atomic and molecular energies on the electrostatic potentials at the nuclei.

This peculiar approach [9,61,79] with no overlap or cross-terms, made the present bond energy theory possible.

In contrast, schemes that result in "nonatomiclike" descriptions meet with the usual difficulties that are encountered when it comes down to fairly distributing interelectronic and internuclear repulsion terms among chemical bonds or "atoms in the molecule"; these terms are not simply separable into atomic or bond contributions. And yet, with the help of the Thomas–Fermi-like approximation $E^v = (1/\gamma^v)(V_{ne}^{eff} + 2V_{nn}^{eff})$, we avoid this sort of partitioning problem, as E^v is expressed solely in terms of the effective potentials at the individual nuclei, which raise no partitioning-related problems. The new problem—that brought up by $1/\gamma^v$—can be solved (in principle) by remembering that this parameter is a weighted average of atomic $1/\gamma_k^v$ terms that can be treated, at least to a good approximation, as constants for each type of atom k.

In short, the core–valence partitioning in real space offers the great advantage of being naturally best suited in problems concerned with real-space atom-by-atom decompositions of molecules. Yet, although serving different purposes, and however different they may seem, real-space and orbital-space core–valence separations appear for what they are: two facets of the same reality. The route to this result was not overly exciting, I am afraid, but the final result certainly justifies our patience.

The role of atomic charges—or, more precisely, of charge variations—manifests itself in the term $\Delta(V_{ne}^{eff} + 2V_{nn}^{eff})$ that will be used in the derivation of the bond energy formulas. Now, the main object of this book—the description of bond energies—is more than a declared goal; it turns out to represent a formidable simplification in the utilization of the $\Delta(V_{ne}^{eff} + 2V_{nn}^{eff})$ term. The nuclear–electronic potential energy $V_{ne,k}^{eff}$ (or its variation $\Delta V_{ne,k}^{eff}$) at nucleus k depends on the valence electrons associated with all the nuclei found in a molecule. In the study of bond energies, however, only the contribution of atom k and those of atoms l bonded to k are retained in the evaluation of the kl bond energies, while all contributions of the more distant centers, not bonded to k or l, are collected in what will finally be treated as "nonbonded interactions." The simplification thus introduced by our bond energy approach is valid; it covers over 99.9% of all interactions that keep the atoms united in a molecule. Hence our interest in local atomic charges—the energy of the bond formed by atoms k and l depends on their charges.

Let us now examine the nature of the appropriate atomic charges. The notion of atomic charge evidently carries the idea of a mental subdivision of the molecule into atomic regions, but without altering our basic approach—the nature of the problem (and of its solution) is still a matter of electrostatic potentials at the nuclei.

The definition of meaningful atomic charges is certainly not without difficulty. But it does not involve the definitely more severe ones linked to a fair apportioning of interaction energies between subspaces among the same subspaces. This is to our advantage, because the distribution of all interatomic interaction energies, as required in "nonatomiclike" descriptions, is considerably more difficult than the sole distribution of electronic charge.

Hence the approach privileged here, based on Eq. (4.47), that uses the variations of effective nuclear–electronic potential energies ΔV_{ne}^{eff} dictated by local atomic electron populations. Regarding the latter, they are deduced along classical lines, rooted in molecular wavefunctions that determine the electron density at any given point in space.

CHAPTER 5

INDUCTIVE EFFECTS; ATOMIC CHARGES

5.1 INTRODUCTION

One of the most popular concepts in chemistry is that of charge distributions in molecules. Pictorial presentations of charge densities can be offered in a number of ways. In the familiar contour map type [24], for example, contours corresponding to various values of the charge density (or of its difference with respect to the super-position of the free atoms) are plotted for different points in a specified plane of the molecule [21,23,92–95]. This type of presentation is certainly a realistic one, namely, with electron densities calculated from Hartree–Fock wavefunctions, which are generally known to be reasonably accurate. Contour maps alone, however, do not tell us how much charge can be assigned in a meaningful way to the individual atoms of a molecule. Hence our problem: selection of the right atomic charges.

Many attempts have been made, but the numbers do not seem to come out right, assuming that one knows what "right" means in this context. As long as atomic charges are taken as byproducts of molecular calculations and used as an interpretative adjuvant in semiquantitative discussions of chemical problems, the lack of well-justified numerical results may be perceived as an annoyance but does not represent an acute problem in itself. In that vein, arguments in defense of calculated charges often invoke their (at least rough) agreement with the chemist's familiar inductive effects, but tell little about the quality of the results.

5.2 THE INDUCTIVE EFFECTS

The chemist's ideas pertaining to changes of electron densities induced by substitutions include the familiar order of electron-releasing ability of alkyl groups, that is, the inductive order:

$$CH_3 < CH_3CH_2 < (CH_3)_2CH < \cdots < (CH_3)_3C$$

In general, calculated charges are known to agree with the usually accepted variations of inductive effects, although arguments of this sort were often kept at a very prudent, noncommittal semiquantitative level. So far, so good, meaning that we could turn things the other way around; since theoretical charges seem to fill expectations on the basis of the inductive order, we might as well start with the latter and examine what kind of charges would fit the expectations.

Long ago, we have indeed carried the argument one decisive step further, also giving great attention to the quality of the charge distributions used for comparison [27,96]. Correlations involving experimental (kinetic and equilibrium) data clearly suggest a linear numerical ordering of the inductive effects. In Taft's scale of polar σ^* constants (Table 5.1), the σ^* parameters are increasingly negative as the groups they describe are better electron donors [96,97].

Our analysis considers an alkane as an alkyl group R attached to either CH_3 or H. Tentatively assuming the validity of the current interpretation of the inductive effects described by Taft's scale, we write

$$q_{CH_3} = a\sigma_R^* \tag{5.1}$$

$$q_H = a\sigma_R^* + b \tag{5.2}$$

Equation (5.1) applies for alkanes described as $R—CH_3$, and q_{CH_3} is the net charge of a methyl group attached to R. Similarly, Eq. (5.2) pertains to alkanes written as $R—H$ and q_H is the net charge of H. The validity of these equations is convincingly demonstrated by SCF Mulliken net charges: Mulliken SCF charges given by every basis set satisfy Eqs. (5.1) and (5.2) [27,44,96]. (See Fig. 5.1.)

Thus we could use the charges given by any LCAO basis of our choice to create an arbitrary set of scaling parameters (as a replacement for the σ^* constants) and proceed with Eqs. (5.1) and (5.2) without invoking Taft's inductive effects. For convenience, however, we shall go along with the σ^* constants of Table 5.1.

Atomic charges obtained from different bases are very bewildering. Equations (5.1) and (5.2) epitomize what they have in common and the reason why they

TABLE 5.1. Polar σ^* Constants

Group R	$\sigma^*(R)$	Group R	$\sigma^*(R)$
CH_3	0	$neo\text{-}C_5H_{11}$	-0.151
C_2H_5	-0.100	$iso\text{-}C_3H_7$	-0.190
$n\text{-}C_3H_7$	-0.115	$sec\text{-}C_4H_9$	-0.210
$n\text{-}C_4H_9$	-0.124	$(C_2H_5)_2CH$	-0.225
$i\text{-}C_4H_9$	-0.129	$tert\text{-}C_4H_9$	-0.300

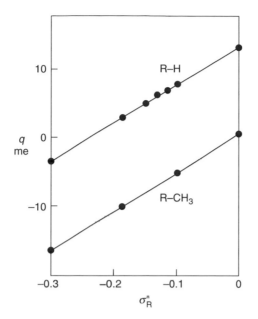

Figure 5.1. Verification of Eqs. (5.1) and (5.2) by means of Mulliken charges deduced from STO-3G calculations involving complete optimizations of geometry and orbital exponents. The net charge q is q_{CH_3} of (5.1) for the lower line and q_H of Eq. (5.2) for R—H [44].

differ: discussion of their basis set dependence now comes down to discussing the parameters a and b. This is where we can turn things to our advantage.

The idea is simple. Since SCF Mulliken charges satisfy Eqs. (5.1) and (5.2), we reverse the argument and deduce an internally consistent set of charges starting off with (5.1) and (5.2) and the σ^* constants of Table 5.1. Charge normalization

$$\sum_k q_k = 0 \qquad (5.3)$$

is part of this calculation. Now we proceed with Eqs.(5.1) and (5.2), using Taft's polar σ^* constants.

The general outline of this backcalculation of charges can be illustrated as follows, taking propane as an example. In this molecule there are four unknown charges: those of the primary and secondary carbon atoms and of two different hydrogens (the weighted average of the primary H atoms is considered in this case). Because of (5.3), three equations are required for solving the problem of the three remaining unknowns. One of these equations is (5.1), considering propane as C_2H_5—CH_3; the other two are given by (5.2), considering propane as n-C_3H_7—H (for the calculation of the primary H atoms) or as iso-C_3H_7—H (for the secondary H atoms). Hence we can solve the problem.

Of course, this approach is by no means a general way of obtaining charges, but the fact that it can be applied to an adequate collection of molecules is sufficient for our intended purpose.

To begin with, let the carbon net charge of ethane $= 1$ arbitrary unit and express the other charges with respect to that reference. Then write

$$a = -\frac{10}{3n} \tag{5.4}$$

because this expression for the slope a in terms of a new variable n is convenient in the presentation of the final results. Eq. (5.1) thus becomes

$$q_{CH_3} = -\left(\frac{10}{3n}\right)\sigma_R^* \tag{5.5}$$

On the other hand, (5.2) shows that the ethane-H net charge, $-\frac{1}{3}q_C^{C_2H_6}$, is $-\frac{1}{3} = b - 0.1a$ (in arbitrary units); hence we get from (5.4) that

$$b = -\frac{n+1}{3n} \tag{5.6}$$

We know from (5.2) that b is the hydrogen net charge in methane because $\sigma^* = 0$ for $R = CH_3$. Using (5.4) and (5.6), we can rewrite (5.2) as follows

$$q_H = -\frac{10\sigma_R^* + n + 1}{3n} \tag{5.7}$$

Finally, Eqs. (5.3), (5.5), (5.7), and the σ_R^* constants of Table 5.1 give the results listed in Table 5.2 by reference to the unit charge defined by $q_C^{C_2H_6} = 1$.

TABLE 5.2. Net Charges (Relative Units)

Molecule	Atom	Net Charge
Methane	C	$4(n+1)/3n$
Ethane	C	1.000
Propane	C_{prim}	$(3n + 0.55)/3n$
	C_{sec}	$(2n - 3.8)/3n$
	H_{prim}	$(0.15 - n)/3n$
	H_{sec}	$(0.9 - n)/3n$
Butane	C_{prim}	$(3n + 0.43)/3n$
	C_{sec}	$(2n - 3.35)/3n$
	H_{prim}	$(0.24 - n)/3n$
	H_{sec}	$(1.1 - n)/3n$
Pentane	C_{centr}	$(2n - 2.8)/3n$
	H_{centr}	$(1.25 - n)/3n$
Isobutane	C_{prim}	$(3n + 1.03)/3n$
	C_{tert}	$(n - 7.7)/3n$
	H_{prim}	$(0.29 - n)/3n$
	H_{tert}	$(2 - n)/3n$
Neopentane	C_{prim}	$(n + 0.49)/n$
	C_{quat}	$-4/n$
	H	$(0.51 - n)/3n$

Table 5.2 is a compact presentation of charge results given by any LCAO-MO method. Tests are straightforward. The quality of the agreement depends on the precision of calculated SCF charges. Results are consistently improved by careful geometry optimizations and, most importantly, when all exponents, including those of the carbon K shells, are optimized individually for each molecule and each non-equivalent atom in the same molecule until stable charges (say, within ± 0.01 me) are obtained [38,98]. Practically, this is feasible only with small basis sets. On the other hand, the requirement for this sort of detailed ζ-optimizations is somewhat less stringent with the use of extended basis sets.

The following example is worked out for Mulliken charges calculated from optimized STO-3G wavefunctions. Using the C charges of methane and ethane, -42.92 and -20.96 me, respectively, $(-48.92)/(-20.96)$ is $4(n + 1)/3n$, which gives $n = 1.3325$. We apply this n in Table 5.2 and obtain the charges in relative units, which must be multiplied by -20.96 to give the results in me (millielectron) units (Table 5.3). Tests of this sort are equally conclusive for semiempirical (e.g., INDO [99], PCILO [100], and extended Hückel [101]) as for large Gaussian basis set calculations [27,96]. What differs from one method to another are n and the ethane carbon net charge. A GTO($6s$ $3p|3s$) basis, for example, leads to $n = 14.11$ and $q_C^{C_2H_6} = -232.5$ me [102], whereas Leroy's ($7s$ $3p|3s$) calculations [103] correspond to $n = 42.3$ and $q_C^{C_2H_6} = -573$ me.

Comparisons between STO-3G, GTO ($6s$ $3p | 3s$), and SCF-Xα-SW net charges with those of Table 5.2 are presented in Tables 5.3 and 5.4.

The tests presented in Tables 5.3 and 5.4—and many others of the same kind [27]—are certainly convincing. If we feel confortable with population analyses that mirror our understanding of electron-releasing (or -withdrawing) abilities of alkyl groups, we should learn the lesson: bluntly stated, we should reject charge analyses that fail to agree with the inductive order.

This brings up the obvious question: What are the "true" values of n and of the net carbon charge in ethane?

TABLE 5.3. Optimized STO-3G Mulliken Net Charges and Charges Deduced for $n = 1.3325$

Molecule	Atom	STO-3G	$n = 1.3325$
Methane	C	-48.92	-48.92
Ethane	C	-20.96	-20.96
Propane	C_{prim}	-23.81	-23.84
	C_{sec}	5.94	5.95
	H_{prim}	6.23	6.20
	H_{sec}	2.20	2.27
Isobutane	C_{prim}	-26.39	-26.36
	C_{tert}	33.36	33.39
	H_{prim}	5.50	5.47
	H_{tert}	-3.53	-3.50
Neopentane	C_{prim}	-28.66	-28.67
	C_{quat}	62.92	62.92

TABLE 5.4. Theoretical Net Atomic Charges^a and Charges Deduced from Formulas in Table 5.2 (au)

Molecule	Atom	GTO(6s 3p|3s)	$n = 14.11$	SCF-Xα-SW	$n = -4.4293$
Methane	C	−0.335	−0.332	0.630	0.632
Ethane	C	−0.234	−0.233	0.610	0.612
Propane	C_{prim}	−0.237	−0.236	0.595	0.587
	C_{sec}	−0.133	−0.135	0.591	0.583
	H_{prim} ^{b}	0.076	0.077	−0.214	−0.211
	H_{sec}	0.075	0.073	−0.247	−0.246
Isobutane	C_{prim}	−0.238	−0.238	0.561	0.565
	C_{tert}	−0.028	−0.036	0.557	0.559
	H_{prim} ^{b}	0.075	0.076	−0.216	−0.217
	H_{tert}	0.072	0.067	−0.295	−0.296
Neopentane	C_{prim}	−0.237	−0.240	0.539	0.544
	C_{quat}	0.060	0.064	0.550	0.553
	H	0.074	0.075	−0.226	−0.228

^{a}The GTO(6s 3p|3s) results are from Ref. 102; the SCF-X-α-SW results are from Ref. 104. The SW-Xα population analysis is discussed in Chapter 8.
^{b}Weighted average of nonequivalent H atoms.

5.3 MEANINGFUL ATOMIC CHARGES

The meaning of n can be inferred from Eq. (5.4) where a measures, in a way, the sensitivity of charge variations to substituent effects. Small $|n|$ values indicate strong substituent effects. If inductive effects did no exist, the charges would be those corresponding to $|n| = \infty$ (i.e., $a = 0$), and all H atoms would carry the same charge. No theoretical method leads to this extreme result.

The sign of a is of utmost importance. Equation (5.2) shows that a must be positive in order to reflect the usual order of electron-releasing abilities $tert$-C$_4$H$_9 > \cdots >$ CH$_3$ because only then will a hydrogen atom attached to a $tert$-butyl group be electron-richer than that of methane. Similarly, as indicated by Eq. (5.1), only then will the methyl group in propane carry a net negative charge, which is an important constraint; pertinent experimental evidence for the $(CH_3)^- - (C_2H_5)^+$ polarity is offered in Ref. 105. Equation (5.4) expresses a in arbitrary units. The corresponding expression in charge units is

$$a = -\left(\frac{10}{3n}\right) q_C^{C_2H_6} \tag{5.8}$$

so that $q_C^{C_2H_6}$ and n must be of opposite signs in order to satisfy the constraint $a > 0$. Ab initio Mulliken charges usually[1] correspond to $n > 0$ and $q_C^{C_2H_6} < 0$, as is the case with most semiempirical results; INDO is the notable exception with $n = -2$

[1]Addition of diffuse functions to the basis may result in $n < 0$ and $q_C^{C_2H_6} > 0$ [106].

and $q_c^{C_2H_6} = +71$ me [27]. It is worth remembering that *any LCAO-MO method leading to an* n *value with the same sign as the carbon net charge of ethane is bound to describe charge variations in the wrong order.* There is no point in closing the eyes on this—it would only make things darker. The test is easy; it suffices to calculate methane and ethane and get *n*.

Charge–NMR Shift Correlations

Our interest in correlations between NMR shifts and atomic electron populations originates in the quest of practical means for obtaining atomic charges. Chapter 6 covers this subject in detail, but there is one aspect of immediate interest. A most instructive comparison using the STO-3G Mulliken net charges of alkanes (Table 5.3) is shown in Fig. 5.2. The points for the CH_3, CH_2, and CH carbon atoms, including also those of cyclohexane [107] and adamantane [39], lie on parallel, equidistant lines shifted from one another by $\sim$30 me. This figure suggests how the three lines can be made to merge into one single correlation line. Indeed, considering a linear relationship for the chemical shifts δ, namely, $\delta = \alpha \times q_C + \delta°$, we can rewrite it as follows

$$\delta = \alpha\left(q_C^{Mull} + N_{CH}\,p\right) + \delta° \qquad (5.9)$$

where N_{CH} is the number of hydrogen atoms attached to a given carbon atom and p stands for the correction required for each hydrogen attached to that carbon, so that

$$q_C = q_C^{Mull} + N_{CH}\,p \qquad (5.10)$$

represents the appropriately "corrected" atomic charges of the carbons.

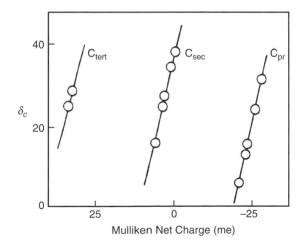

Figure 5.2. Comparison between the carbon-13 NMR shifts (ppm from TMS) of selected primary, secondary, and tertiary sp^3 carbon atoms and Mulliken net atomic charges (in me units).

The corresponding correction for hydrogen is as follows, from obvious charge normalization considerations:

$$q_H = q_H^{Mull} - p \tag{5.11}$$

The parameter p is deduced from a multiple regression analysis using Mulliken charges and the appropriate experimental NMR shifts. For the STO-3G charges used in this example, it is $p = 30.12$ me.

Energy calculations [108], on the other hand, offer an empirical but quite reliable evaluation: $q_C^{C_2H_6} \approx 35.1$ me, which is henceforth adopted. Similar considerations apply to sp^2 carbons as well.

The comparison between the net atomic charges (q_C^{Mull}), deduced from optimized STO-3G computations, and the experimental shifts of olefinic sp^2 carbon atoms [40] yields a result similar to that observed for sp^3 carbons; again we have three equidistant parallel regression lines, depending on the number of H atoms attached to the olefinic carbon. The analysis by means of Eq. (5.9), on the other hand, leads to $p = 31.8 \pm 4$ me, that is, in essence, the value found for the sp^3 carbons, 30.12 me [40]. Energy calculations [109] led to the empirical result $q_C^{C_2H_4} = 7.7$ me for the ethylene carbon atom.

The bottom line is that comparisons of SCF Mulliken net charges with NMR shifts suggest that the alkane and alkene carbon and hydrogen net charges should be corrected as shown in Eqs. (5.10) and (5.11), respectively, using the same p value.

Auxiliary Relationships

We can now ask how the modified charges q_C and q_H of Eqs. (5.10) and (5.11) compare with their original (Mulliken) counterparts. Let us define $q_C^{C_2H_6,Mull} =$ ethane-C Mulliken net charge, with $n = n^{Mull}$ and $q_C^{C_2H_6} =$ modified ethane-C net charge, corresponding to $n = n$. Equations (5.8) and (5.11) tell us that the slopes of the hydrogen net charges versus σ^* are $-10q_C^{C_2H_6}/3n$ for the modified charges and $-10q_C^{C_2H_6,Mull}/3n^{Mull}$ for the original Mulliken charges. It is clear, however, that the transformation shown in Eq. (5.11) leaves the slope of the hydrogen net charge versus σ^* unaffected, thus indicating that

$$\frac{q_C^{C_2H_6}}{n} = \frac{q_C^{C_2H_6,Mull}}{n^{Mull}} \tag{5.12}$$

Finally, according to Eq. (5.10), we have

$$p = \frac{1}{3}\left(q_C^{C_2H_6} - q_C^{C_2H_6,Mull}\right) \tag{5.13}$$

$$p = \frac{n - n^{Mull}}{3n^{Mull}} q_C^{C_2H_6,Mull} \tag{5.14}$$

For example, the charge–NMR shift correlation [Eq.(5.9)] gives $n = -4.4122$ for $p = 30.12$ me. Using this result and the formula given for the methane carbon atom (Table 5.2), we get the useful formula

$$p = \frac{3.4122}{3} q_C^{C_2H_6,Mull} - \frac{4.4122}{4} q_C^{CH_4,Mull} \tag{5.15}$$

which facilitates the proper rescaling of Mulliken net charges.

Selection of n

There is one particular n that merits special attention. It reflects simple customary ideas: "The electron-attracting power of otherwise similar atoms decreases as their electron populations increase, thus opposing charge separation." This concept views local charge variations as events occurring "most reluctantly," suggesting that the carbon atoms found in alkanes should be very similar to one another, conceivably differing as little as possible from one another.

For a set of alkanes, each one containing two different carbons with net charges q_r and q_s, this constraint amounts to minimizing the sum $\sum (q_r - q_s)^2$ over the set. Using the formulas of Table 5.2, it is found that

$$\frac{d \sum (q_r - q_s)^2}{dn} = 0$$

is satified by $n \approx -4.4$. Of course, this result for n does not depend on the particular set of charges used in this type of calculation; only p is basis-set-dependent, not the n deduced from this minimization of charge variations. Empirical evaluations of n (like that reported for NMR shifts) consistently show that n is of the order of ~ -4.4.

The negative n value means that the alkane carbon net charges are positive. This relatively important C^+–H^- polarity is in line with the view that hydrogen is certainly more electronegative than carbon, as Mulliken and Roothaan [110] and others [111–113] have pointed out.

Now it is a consequence of postulating minimal charge variations.

5.4 SELECTED REFERENCE NET ATOMIC CHARGES

Here we examine the carbon net charges of ethane and ethylene, obtained from SCF and configuration interaction calculations, corrected by means of the appropriate p, determined for $n = -4.4122$. Remember that the same value of p applies to both ethane and ethylene, as n is solely determined by the effectiveness of the inductive effects. Equation (5.15) is used to get p, namely, $p = 138.68$ me in 4-31G + CI calculations and thus, from Eq. (5.10), the corresponding carbon charges of ethane and ethylene (see Table 5.5).

TABLE 5.5. Mulliken Net Charges and Final Net Charges [Eq. (5.10)] (me)

	Mulliken Net Charge of Carbon			Final C Net Charge	
Calculation	Methane	Ethane	Ethylene	Ethane	Ethylene
STO-3G	−48.92	−20.96	−128.4	69.40	−68.2
+CI	−46.37	−22.28	—	55.14	—
4-31G	−523.2	−382.7	−346.4	42.8	−62.7
+CI	−515.7	−378.2	−269.9	37.8	7.5
Empirical	—	—	—	35.1	7.7

The SDCI calculations are self-explanatory; a glance at the results reveals that the 4-31G + CI charges [51] are indeed remarkably close to their empirical counterparts, which are optimum values deduced from statistical best fits in energy calculations [108,109].

No such SDCI results are presently available for benzene, but taking advantage of the fact that the carbon atoms are evidently electroneutral in graphite and not so in benzenoid hydrocarbons, the results obtained for graphite support the approximate validity of bond energies deduced for polynuclear benzenoid hydrocarbons and of the net charge, 13.2 me (probably ±1 me), deduced for the carbon atom of benzene [44].

Selected empirical "best fit" reference charges are reported in Table 5.6. The results for nitrogen and oxygen are described in Part III, as are the relevant, highly accurate charge–NMR shift correlations.

In summary, the charges suited for our energy calculations are clearly identified with those that correlate to NMR chemical shifts. They closely reflect the expected familiar inductive effects. This observation should not come as a surprise if we think of atomic charges as part of the description of nature at the molecular level and hence reject the idea that each phenomenon or property demands its very own definition of atomic charge. But it must also be made clear that we think here first and foremost of *point charges* allowing the interpretation of properties for which they are an admissible simplified representation of integrated charge densities. If not, our picture of point charges no longer holds.

TABLE 5.6. Selected Reference Net Charges

Atom k	Host	q_k (me)	Reference
C	C_2H_6	35.1	Chapter 13 [108]
	C_2H_4	7.7	Chapter 14 [109]
	C_6H_6	13.2	Chapter 14 [44]
N	CH_3NH_2	−9.00	Chapter 15 [34]
O	$(C_2H_5)_2O$	5.18	Chapter 16 [44]
	CH_3OH	10.53	Chapter 16

As it is, it should be remembered that Fig. 5.2 has become a sort of Rosetta stone whose deciphering paves the way toward a better understanding of what falsely seems to be an inherent intricacy of Mulliken's population analysis.

But before getting there, let us discuss charge–shift correlations, as well as another instructive topic, one that defines charges in light of measured adiabatic ionization potentials.

CHAPTER 6

ATOMIC CHARGES AND NMR SHIFTS

6.1 SCOPE

Carefully established correlations between nuclear magnetic resonance (NMR) shifts and atomic electron populations in well-defined series of closely related compounds can prove valuable for the evaluation of atomic charges in similar systems that are at, or beyond, the limits of practical computational feasibility. We certainly could make good use of them. [Also remember the insight gained with the help of Fig. 5.2; it led to Eq. (5.10).]

This sort of approach postulates that one of the major factors governing the shielding of a specific nucleus is its local electron density. Now, chemical shift is a property of the interaction of the charge density with an external magnetic field. It depends therefore on the value of the integrated charge density (or "charge") in the neighborhood of a nucleus (as well as on other factors, of course, e.g., the magnetic susceptibility of that charge density), but a formal relationship between NMR chemical shifts and atomic charges is not part of the rigorous theory of nuclear magnetic resonance.

Under these circumstances, it seems preferable to develop arguments in favor of charge–shift correlations from within the theory of magnetic shielding with no reference to any particular population analysis, rather than proceeding with brute-force attempts at correlating observed NMR shifts with atomic charges, assuming that one knows how to define them.

Atomic Charges, Bond Properties, and Molecular Energies, by Sándor Fliszár
Copyright © 2009 John Wiley & Sons, Inc.

6.2 INTRODUCTION

Following the convention adopted for ^{13}C and ^{17}O (but not for ^{15}N) nuclei, in writing a linear relationship

$$\delta = \alpha \times q + \delta^\circ \tag{6.1}$$

between chemical shifts δ and net (i.e., nuclear minus electronic) charges q, we must keep in mind that

- Increasingly positive δ values correspond to downfield shifts.
- q becomes more negative as the corresponding electron population increases.

Hence, a positive slope α indicates that an increase of electronic charge at an atom results in a high-field shift, reflected by a lowering of δ. Conversely, a negative slope α indicates that an increase in local electron population (more negative q) results in a downfield shift. The puzzling point is that both positive and negative slopes are met in applications of Eq. (6.1), such as $\alpha > 0$ for ethylenic carbon [40] and carbonyl oxygen atoms [41] and $\alpha < 0$ for paraffinic carbon [38] and ether oxygen atoms [41]. This is a problem well worth looking into. Before doing so, however, let us examine a few general aspects regarding the postulated validity of charge–shift relationships.

The main conceptual difficulty stems from the fact that the attemps at correlating NMR shifts with atomic electron populations are rooted in one's intuition rather than being based on a rigorous formalism.

6.3 MERITS OF CHARGE–SHIFT RELATIONSHIPS

Fortunately, we can take advantage of an indirect way of assessing the merits of charge–shift correlations by examining the average diamagnetic and paramagnetic contributions, σ^d and σ^p, respectively, to the total average magnetic shielding:

$$\sigma = \sigma^d + \sigma^p$$

The rigorous theory is well known [114], but approximations for σ^d and σ^p are used in the following discussion of results derived by means of the formalism given by Vauthier et al. [115].

This approach, based on Pople's finite perturbation theory [116], involves the INDO approximations [117] on a gauge-invariant GIAO basis [118] and London's approximation. Moreover, it satisfies the Hermitian requirement for the first-order perturbation matrix reflecting the effect of an applied external magnetic field. The latter condition results in a significant improvement of calculated ^{13}C magnetic shieldings, the average precision being ~ 5 ppm [115].

The point is that this formalism for σ permits a separation into mono-, di-, and triatomic contributions, thus revealing the relative weight of "local" and "distant" electron densities in the magnetic shielding of a given nucleus. In this manner, it becomes possible to gain a reasonable estimate about the chances that chemicals shifts do, indeed, depend primarily on local electronic populations, at least in series of closely related compounds. The most detailed results are those derived for ethylenic and acetylenic sp^2 and sp carbon atoms, respectively.

To begin with, it appears that the local diamagnetic contribution to the magnetic shielding is practically the same for all sp^3, sp^2, and sp carbon nuclei (57.85 ± 0.6 ppm). Moreover, the results for sp^2 carbons indicate that the *total* diamagnetic part (including all contributions from distant atoms) *plus* the paramagnetic part due to the distant atoms is nearly constant (82.7 ppm), within ~ 0.4 ppm. The gap between this sum and the total magnetic shielding represents the paramagnetic contribution excluding that of distant atoms, that is, the local paramagnetic shielding plus the paramagnetic part contributed by the neighbors of the nucleus under scrutiny. It is this gap that reflects the total variation in magnetic shielding (or, at least, its major part by far) for a given nucleus in a series of closely related compounds; it is now at the center of our attention. The effects of the neighboring atoms that are included in this paramagnetic shielding are listed in Table 6.1. The results reflect the smallness of these effects.

For ethylenic and acetylenic carbon atoms, one can consider the neighbors' contributions as being constant, or nearly so (within ~ 1.5 ppm). The corresponding uncertainty introduced by assuming constant neighbors' contributions for sp^3 carbon atoms probably does not exceed ~ 0.3 ppm. As a consequence, in a series of closely related compounds, the variations of the local paramagnetic shielding appear to represent the largest part, by far, of the total changes in shielding experienced by a given nucleus, for example, by sp^2 carbons in a series of ethylenes or sp^3-hybridized carbon atoms in paraffins.

Similar conclusions are also reached for the magnetic shielding of ^{15}N atoms, as revealed by a detailed study of a series of amines, nitriles, ammonia, pyridine, pyrazine, pyrimidine, and pyridazine [119]. Their local diamagnetic shielding is virtually

TABLE 6.1. Paramagnetic Shielding Contributed by Neighboring Atoms (ppm)[a]

Molecule (Atom*)	Shielding	Molecule (Atom*)	Shielding
C^*H_4	0.17	$(CH_3)_2C=C^*HCH_3$	0.38
$CH_3C^*H_3$	-0.11	$(CH_3)_2C^*=CH_2$	-1.15
$CH_2=C^*H_2$	-1.21	$(CH_3)_2C^*=CHCH_3$	1.22
$CH_3CH=C^*H_2$	0.70	$(CH_3)_2C^*=C(CH_3)_2$	1.52
$(CH_3)_2C=C^*H_2$	1.52	$CH\equiv C^*H$	1.65
$CH_3C^*H=CH_2$	-2.15	$CH_3C\equiv C^*H$	3.79
$CH_3C^*H=CHCH_3$ *cis*	0.62	$CH_3C^*\equiv CH$	1.15
$CH_3C^*H=CHCH_3$ *trans*	0.67	$CH_3C^*\equiv CCH_3$	3.09

[a]These results were deduced from those indicated in Ref. 115 and represent $\sigma^P(KK) + \sigma^P_z(MK)$, as defined in this reference. The local paramagnetic shielding discussed in the text is $\sigma^P(M)$ (Eq. 9 of Ref. 115).
Source: Ref. 115.

constant ($\sim$88.2 $\pm$ 1.2 ppm), and so are the sums of all the contributions other than the local (dia- and paramagnetic) ones ($\sim$33.1 $\pm$ 4 ppm), in a range of $\sim$400 ppm.

Therefore, within the precision of the present type of analysis, it seems quite reasonable to anticipate correlations between NMR shifts and atomic charges, which, of course, are strictly local properties. Following this analysis of the individual nonlocal effects revealing, namely, the small participation of tricentric integrals involving distant atoms, the practical validity of charge–shift relationships rests largely with cancellation effects of a number of terms that, to begin with, are small or relatively constant. The importance of the nonlocal contributions is further reduced with the selection of a scale tailored for comparisons between atoms of the same type, with reference to an appropriately chosen member of that series. (The fact that "distant" atoms have only a minor, if any, effect on the shielding of heavy nuclei justifies in part the solvaton model [120] discussed by Jallah-Heravi and Webb [121] and the smallness of solvent effects on ^{13}C shifts.) For hydrogen atoms, however, the situation is different because of the large weight of the three-center integrals in the calculation of their magnetic shielding.

So far we have learned that, in certain series of related compounds, it is the local paramagnetic shielding that governs the changes in total shielding

$$\Delta\sigma_{\text{total}} \simeq \Delta\sigma_{\text{local}}^{\text{p}} \qquad (6.2)$$

and, hence, that under these circumstances it may well be justified to expect correlations between NMR shifts and local electron populations. It remains, however, that charge–shift correlations are essentially empirical in nature; while the definition of "closely related compounds" may be linked to the approximate validity of Eq. (6.2), the practical answer stems ultimately from the actual examination of shift–charge results.

6.4   AROMATIC HYDROCARBONS

To begin with, let us examine the probably most quoted plot, that of the familiar Spiesecke–Schneider work [122] relating the ^{13}C NMR shifts of tropylium ion, benzene, cyclopentadienyl anion, and cyclooctatetraene dianion to the corresponding carbon atomic charges. The latter were deduced by assuming the local π-electron density to be known from the number of π electrons and the number of carbon atoms over which the π cloud was distributed. The estimated shift, $\sim$160 ppm per electron, has become an almost unerasable part of our grammar. The linear correlation between ^{13}C chemical shifts and π charge density was later extended to 2π electron systems [123–125] as well as to the 10π cyclononatetraene anion [126]. A plot of this correlation for the whole series was presented by Olah and Mateescu [123], who used, where appropriate, simple Hückel molecular orbital theory for deducing charge distributions.

At a quite different level of approximation, this class of compounds was investigated by means of STO-3G calculations involving a detailed optimization of all the geometric and ζ exponent parameters [42]. The Mulliken net atomic charges and the

TABLE 6.2. Carbon Net Charges and NMR Shifts of Selected Aromatic Hydrocarbons[a] (me)

Compound	q_σ	q_π	q_{total}	δ^b
1 Cyclopropenium cation, $C_3H_3^+$	−223.5	333.3	109.8	176.8
2 Cycloheptatriene cation, $C_7H_7^+$	−120.5	142.9	22.4	155.4
3 Benzene, C_6H_6	−47.4	0	−47.4	128.7
4 Cyclononatetraenide anion, $C_9H_9^-$	5.1	−111.1	−106.0	108.8
5 Cyclopentadienide anion, $C_5H_5^-$	26.5	−200.0	−173.5	102.1
6 Cyclooctatetraenide dianion, $C_8H_8^{2-}$	75.0	−250.0	−175.0	85.3

[a]Results expressed relative to **1**, **5**, and **6** electrons, respectively, for the π, σ, and total net charges. A negative sign indicates an increase in electron population.
[b]Specifically, ppm, from Me_4Si.
Source: Ref. 42.

corresponding chemical shifts (Table 6.2) yield the correlation presented in Fig. 6.1. In spite of some scatter about the regression line, a point that is discussed below, it appears that Eq. (6.1) is reasonably well satisfied with the use of total ($\sigma + \pi$) net atomic charges, with $\alpha \simeq 300$ ppm/electron.

The behavior of the *para* carbon atoms of substituted benzenes is similar. Using STO-3G Mulliken $\sigma + \pi$ net charges, the correlation with NMR shifts resembles that shown in Fig. 6.1, with $\alpha \simeq 384$ ppm/e [127].

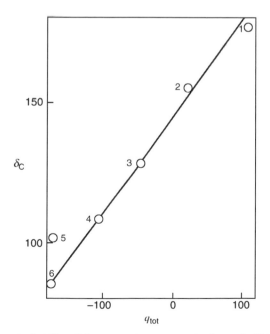

Figure 6.1. ^{13}C chemical shifts of the aromatic compounds shown in Table 6.2 versus total ($\sigma + \pi$) net charges (ppm from TMS, viz., me).

A similar study on *meta* carbons [127], while giving results of the same type, is perhaps somewhat less conclusive because of the very limited range of variation of the *meta*-carbon NMR shifts ($\sim$1.5 ppm). It remains, however, that the major conclusions drawn for the *para* carbons apply to the *meta* carbons as well.

As one would anticipate from the similarity in the chemical nature of these substituted benzenes and the compounds indicated in Table 6.2, the gross features are quite similar, namely, in terms of the increase in electron population at carbon resulting in a high-field shift.

Not too much importance should be given to the difference between the slopes α calculated for the two series of compounds. Part of the difference is possibly due to the fact that the substituted benzenes were calculated using the "standard" STO-3G method, which is certainly a reasonable approach for this class of molecules, whereas the STO-3G remake of the Spiesecke–Schneider correlation has involved extensive geometry and scale factor optimizations, dictated by the diversity of the members of this series. In addition, one should consider that the Spiesecke–Schneider correlation involves cycles of different size, a circumstance that introduces an uncertainty regarding the validity (or lack of it) of interpreting chemical shift differences as a function of Mulliken charge density only, disregarding possible effects linked to its shape. An indication about the overall influence of ring size on the quality of simple charge–shift correlations in this class of compounds revealed [42] that this effect is relatively modest when transposed on the scale of the correlation given in Fig. 6.1.

The charge–shift correlation presented in Fig. 6.1 is, on the whole, reasonably good, mainly because it covers an important range of shift and charge results. The results [127] for the substituted benzenes are more significant because they do not suffer from possible drawbacks linked to ring size. Indeed, their correlation is superior in quality to that given in Fig. 6.1.

Charge–Shift Correlations of Aromatic Hydrocarbons

Traditionally, much of the discussion reported in the literature about ^{13}C NMR shifts and electronic structure has related to aromatic systems, following Lauterbur's suggestion [128] that in these systems the shielding is governed primarily by the π-electron density at the carbon nuclei. Although the analysis presented here has emphasized relationships with total ($\sigma + \pi$) atomic charges, there is no doubt that correlations with π-electron populations have their merit. For example, the ^{13}C shifts of the aromatics described in Table 6.2 yield an excellent linear correlation with π charges, showing that any increase of π-electron population at carbon provokes a high-field shift [42]. But it is also true that an equally good linear correlation is obtained if σ charges are used instead; note, however, that any increase of σ-electron population is accompanied by an important downfield shift. This observation suffices to warn us that for aromatic (and, more generally, sp^2 carbon) systems the evaluation of the dependence of NMR shifts on electronic charges should not be restricted to π electrons only, disregarding σ charges. The study of σ systems would otherwise come to an abrupt end before it has started.

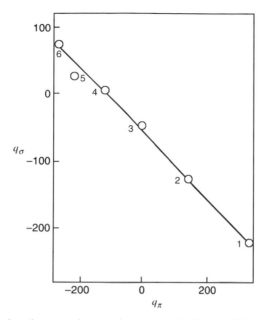

Figure 6.2. Comparison between the σ and π net atomic charges of the aromatic compounds shown in Table 6.2 (me) [42].

The reason why σ, π, and total ($\sigma + \pi$) charges yield correlations of similar quality for the aromatics is due to the linear decrease in σ population accompanying any increase in π electronic charge. Figure 6.2 illustrates this behavior for the compounds described in Table 6.2.

Similarly, in the series of monosubstituted benzenes, the calculated changes in σ and π populations at the *para*-carbon atom are accurately inversly related to one another, as convincingly demonstrated elsewhere [127]. The π population shows the greater change, and the σ population seems to be altered by approximately 55% in the opposite direction. The results obtained for the π and σ populations at the *meta*-carbon atom are similar, but these points show some scatter from linearity. However, most of the *meta* points fall close to the correlation line drawn for the *para*-carbon atoms.

Describing now, where appropriate, the observed changes in σ and π populations by the equation

$$\Delta q_\sigma = m\,\Delta q_\pi \tag{6.3}$$

with $\Delta q = \Delta q_\sigma + \Delta q_\pi$, it appears that Eq. (6.1) can be written as follows

$$\Delta \delta = \alpha_\sigma \Delta q_\sigma + \alpha_\pi \Delta q_\pi \tag{6.4}$$

$$= \frac{m\alpha_\sigma + \alpha_\pi}{m + 1} \times \Delta q \tag{6.5}$$

where $m \alpha_\sigma + \alpha_\pi = (m+1)\alpha$ represents the apparent slope of δ vs. q_π (e.g., ~ 160 ppm/e), which is now seen to also account for the fact that σ and π charges vary in opposite directions. Note that when Eq. (6.3) applies, the individual α_σ and α_π parameters cannot be obtained from simple regression analyses using Eq. (6.4) because Δq_σ and Δq_π are not independent variables.

Formula for the Charge–Shift Correlation

STO-3G charge analyses of benzenoid hydrocarbons [129] indicated that $\delta = 0.835q + 178.66$ ppm from TMS, where q is the calculated net charge of carbon. Taking the σ–π separation into account, Eqs. (6.3) and (6.4), with $m = -0.814$ and $(d\delta/dq_\pi) = 157$ ppm/e, it was thus deduced that

$$q = 1.2\,(\delta - \delta_{C_6H_6}) + 13.2 \text{ me} \tag{6.6}$$

where $\delta_{C_6H_6} = 128.5$ ppm is the chemical shift of benzene, from TMS. This result is admittedly crude. But extensive numerical analyses, such as those reported in energy calculations of benzenoid hydrocarbons using ^{13}C NMR shifts, gave no reason for revision.

6.5   RELATIONSHIPS INVOLVING sp^3 CARBON ATOMS

The validity of Eq. (6.1) has been carefully established for linear and branched paraffins, cyclohexane and methylated cyclohexanes, including molecules consisting of several cyclohexane rings in the chair conformation (e.g., *cis*- and *trans*-decalin, bicyclo[3.3.1]nonane, adamantane, and methylated adamantanes) as well as in boat conformation (e.g., iceane and bicyclo[2.2.2]octane) [44]. No special effect seems to contribute to the chemical shift because of the presence of cyclic structures.

A multiple regression analysis using Eq. (5.9) and carbon charges "corrected" according to Eq. (5.10) leads to [38,44]

$$\delta = -237.1\,\frac{q_C}{q_C^\circ} + 242.64 \text{ ppm from TMS} \tag{6.7}$$

with a standard error of 0.3 ppm. Using ethane as reference, where $q_C^\circ = q_C^{C_2H_6}$ is its carbon net charge, we get

$$\delta - \delta_{C_2H_6} = -237.1\,\frac{(q_C - q_C^{C_2H_6})}{q_C^{C_2H_6}} \text{ ppm from ethane}$$

Note that at this point the carbon charges are expressed in a convenient dimensionless way or, if we prefer, in "relative units," taking the ethane carbon atom as reference by setting its charge at one arbitrary unit.

Now, using familiar charge units, with $q_C^\circ = 35.1$ me (Chapter 5), we deduce that

$$q_C - 35.1 = -0.148\,(\delta - \delta_{C_2H_6}) \text{ me} \tag{6.8}$$

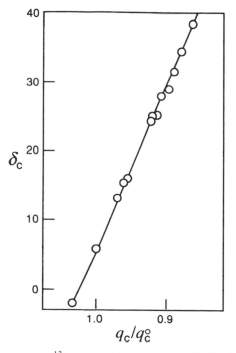

Figure 6.3. Correlation between ^{13}C NMR shifts (ppm from TMS) of sp^3 carbon atoms and net charges from Eq. (5.10) using optimized STO-3G charges with $p = 30.12$ me. The charges are expressed in terms of q_C/q_C°, that is, in relative units, for $n = -4.4122$. This figure now includes the points of Fig. 5.2 plus that of methane and the quaternary carbon of neopentane. (Reproduced with permission from Ref. 44.)

This is probably the most accurate charge–shift correlation presently known. Figure 6.3 illustrates the quality of this correlation.

Incidentally, in sharp contrast with the results obtained for aromatic molecules, note that any increase in electron population at a carbon atom is accompanied by an important downfield shift. This is a result that we shall keep in mind when discussing carbon atoms in typical σ systems.

Equation (6.8) is accurate for acyclic and six-membered cyclic saturated hydrocarbons. Smaller or larger cycles are not described by this equation [44,107,130]. Simple charge–shift correlations fail because of the changes in local geometry affecting the hybridization of carbon.

Del Re and coworkers [131] were concerned with the relation of s character in hybrids to bond angles and have considered hybridization as described by local orbitals, determined by requiring that hybrids on different atoms have minimal overlap unless they participate in the same bond. Alternate approaches are provided by the bond index of Wiberg [132] and by the Trindle–Sinanoğlu procedure [133] for the application of the physical criterion of Lennard-Jones and Pople [134,135], requiring that an electron in a localized orbital interact maximally with the electron sharing that orbital.

A good insight into the problems related to, and the possibilities offered by, the local orbital and bond index characterization of hybridization has been offered by Trindle and Sinanoğlu [136]; when a localized description of the wavefunction is possible—a situation that allows unambiguous definition of hybridization—the two methods give undistinguishable results. Calculated p characters are in good agreement with estimates (rooted in a work by Juan and Gutowsky [137]) derived from NMR coupling constants between carbon-13 nuclei and directly bound protons [132,136].

Theoretical evaluations of hybridization, as well as estimates from NMR coupling constants $J(^{13}CH)$, are anticipated to assist in future work on charge–shift correlations in cases suspected of presenting local geometry changes invalidating simple charge–shift relationships.

A more general form of Eq. (6.8) is

$$\Delta q_C = -0.148 \, \Delta\delta_C \qquad (6.9)$$

where $\Delta q_C = q_C - q_C^{ref.}$ and $\Delta\delta_C = \delta_C - \delta_C^{ref.}$ are defined with respect to appropriately selected reference values, $q_C^{ref.}$ for the carbon charge, corresponding to $\delta_C^{ref.}$ for its chemical shift.

6.6 RELATIONSHIPS INVOLVING OLEFINIC CARBONS

Selected results are presented in Table 6.3 for typical olefinic carbon atoms. The Mulliken net charges were obtained [40] from full (geometry and ζ exponents) optimizations in the STO-3G basis.

TABLE 6.3. Atomic Charges (me) and NMR Shifts[a] of Olefinic Carbon Atoms

Molecule	Atom	$q_{C,\sigma}^{Mull}$	$q_{C,\pi}^{Mull}$	q_C^{total}	$(5.10)^b$	δ
Ethylene	C (**4**)	−128.4	0.0	−128.4	−68.2	122.8
Propene	C-1 (**2**)	−120.4	−34.2	−154.6	−94.4	115.0
	C-2 (**9**)	−83.5	25.2	−58.3	−28.2	133.1
Isobutene	C-1 (**1**)	−112.2	−61.7	−173.9	−113.7	109.8
	C-2 (**10**)	−55.2	47.6	−7.6	−7.6	141.2
trans-Butene	C-2 (**7**)	−78.6	−7.7	−86.3	−56.2	125.8
cis-Butene	C-2 (**6**)	−77.6	−8.8	−86.4	−56.3	124.3
2-Methyl-2-butene	C-2 (**8**)	−49.8	12.3	−37.5	−37.5	131.4
	C-3 (**3**)	−73.8	−32.6	−106.4	−76.3	118.7
2,3-Dimethyl-2-butene	C-2 (**5**)	−50.2	−11.0	−61.2	−61.2	123.2

[a]Values of A. J. Jones and D. M. Grant, reported in Ref. 138.
[b]Corrected carbon charge, Eq. (5.10), using $p = 30.12$ me.
Source: Ref. 40.

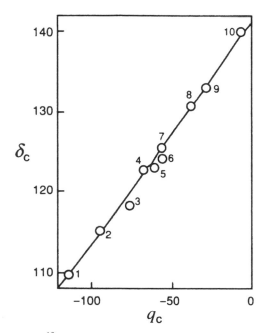

Figure 6.4. Comparison of ^{13}C NMR shifts, ppm from TMS, with the corrected carbon total $(\sigma + \pi)$ net charges [Eq. (5.10)] reported in Table 6.3 for selected olefins. The atom numbering is that indicated in Table 6.3. The radius of the circles represents an uncertainty of 0.7 ppm or 3.5 me. (From Ref. 40.)

The analysis by means of Eq. (5.9) leads to $p = 31.8 \pm 4$ me, which is in essence the value found for the sp^3 carbons, 30.12 me. The correlation between NMR shifts and the SCF carbon charges corrected using $p = 30.12$ me is convincing (Fig. 6.4).

Here again, σ and π populations vary in opposite directions. Using now Eqs. (6.3) and (6.5), that is, the slope

$$\alpha = \frac{m\alpha_\sigma + \alpha_\pi}{m + 1} \tag{6.10}$$

a crude indirect estimate of α can be obtained from $m \simeq -0.955$ and $(d\delta_C/dq_\pi) \simeq 300$ ppm/e, giving

$$q_C - 7.7 \simeq 0.15\,(\delta_C - 122.8) \text{ me} \tag{6.11}$$

This rough estimate is to be taken *cum grano salis* (with a grain of salt)—it follows from energy calculations and brute-force fits with experimental energy data [44]. It is probably not precise, because the contributions of the sp^2-carbon charge variations are rather small and likely to be blurred by uncertainties of the experimental energy data.

6.7   CARBON BONDED TO NITROGEN OR OXYGEN

The charge–shift correlations for carbon atoms bonded to nitrogen or oxygen are described by Eq. (6.9), namely

$$\Delta q_C = -0.148\,\Delta\delta_C$$

with $\Delta q_C = q_C - q_C^{\text{ref.}}$ and $\Delta\delta_C = \delta_C - \delta_C^{\text{ref.}}$.

For the C—N bonds of amines, the analysis presented in Section 15.1 validates this result with the usual reference values $q_C^{\text{ref.}} = 35.1$ me and $\delta_C^{\text{ref.}} = 5.8$ ppm from TMS, which are exactly those of the paraffins. The presence of nitrogen next to carbon does not seem to require any ad hoc revision; this is certainly an acceptable approximation [139], at least within the limits of the energy calculations involved in this evaluation.

Not so when oxygen is the neighboring atom—oxygen introduces an "extra" downfield shift at its bonded α-carbon, estimated [140] at $\sim$41.7 ppm in the case of the ethers. Assuming that this shift, which could be due in part to the electric field of the oxygen dipole, is not primarily a carbon charge effect, we estimate the latter by subtracting 41.7 ppm from the observed α-carbon shifts (60.1, viz., 199.1 ppm from ethane for diethylether and acetone, respectively) giving, with the help of Eq. (6.9), the following reference values for the carbon net charges: $q_C^{\text{ref.}} \simeq 32.4$ me for diethylether and $q_C^{\text{ref.}} \simeq 11.8$ me for acetone.[1] A refinement based on energy calculations [141] led to the following results, to be used with Eq. (6.9):

Ethers: $q_C^{\text{ref.}} = 31.26$ me, $\delta_C^{\text{ref.}} = 65.9$ ppm from TMS
Ketones: $q_C^{\text{ref.}} = 14.0$ me, $\delta_C^{\text{ref.}} = 204.9$ ppm from TMS

The carbons of the alcohols are calculated just like those of the ethers.

Interestingly, the comparison of the carbonyl-^{13}C NMR shifts with their net charges (Table 6.4, Fig. 6.5) indicates the same trend as that observed for the sp^3 carbon atoms in paraffins, namely, a high-field shift with increasing positive net charge.

The slope $\alpha = -0.148$ me/ppm assumed for sp^3 carbons attached to nitrogen is plausible because, at least to a good approximation, these carbons are very similar to those of the paraffins. Regarding the α-carbons of the ethers, the only support in favor of this α comes from energy calculations using it to obtain the required charges; calculated and experimental energies agree within 0.20 kcal/mol (root-mean-square deviation). The selection of $\alpha = -0.148$ me/ppm seems reasonable, though perhaps not very accurate. The same conclusion is reached for the carbonyl α-carbons, although this α seems less credible. It must be added, however, that the

---

[1]In a nutshell: for the diethylether α-carbon, remembering that 60.1 ppm is its shift relative to ethane, the difference $\Delta\delta_C = 60.1 - 47.1$ measures the carbon charge effect, namely, what is due to $q_C - 35.1 = -0.148\,\Delta\delta_C$, which gives, in this approximation, $q_C^{\text{ref.}} = q_C$ (α-C of diethylether) $\simeq 32.4$ me.

**TABLE 6.4. Net Charges (me) and NMR Shifts[a]
of Carbonyl Carbon Atoms**

Molecule	q_C	δ_C
1 $(CH_3)_2CO$	259.7	204.9
2 $CH_3COC_2H_5$	253.6	207.0
3 $CH_3COiC_3H_7$	244.9	210.0
4 $(C_2H_5)_2CO$	247.4	209.4
5 $C_2H_5COi_3H_7$	238.8	212.3
6 $(iC_3H_7)_2CO$	233.8	215.5

[a]Reported in Ref. 142.

Source: Ref. 41.

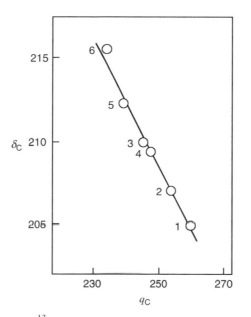

Figure 6.5. Comparison of ^{13}C NMR shifts, ppm from TMS, of ketone carbonyl–carbon atoms with the corresponding atomic charges, reported in Table 6.4. The atom numbering is that indicated in this Table. (Reproduced with permission from Ref. 41.)

contributions of these α-carbons vary so little that they conceal the flaw regarding the validity of the assumed α.

6.8 CORRELATIONS INVOLVING N-15 NMR SHIFTS

This section considers alkylamines, nitroalkanes, isonitriles, pyridines, and diazines and offers comparisons between NMR shifts and calculated atomic charges. Most of the charge results used here were obtained from fully optimized STO-3G calculations [43], but in this peculiar situation it must be made clear that charge analyses are plagued at the outset by an imbalance in the hydrogen basis sets currently used in ab initio calculations [30,143–146], which significantly alters the description of nitrogen itself.

Basis set superposition errors were evaluated [30] by replacing a standard 6-31G* basis by an extended form in which the basis of ammonia H atoms and that of the methyl groups of trimethylamine were retained in the treatment of each alkylamine (i.e., mono-, di-, and trimethylamine) and of ammonia itself. The results indicated that the quality of the treatment of amine nitrogen atoms is strongly dependent on the number of methyl groups because the replacement of alkyl groups by hydrogens greatly impoverishes the basis. This and related studies [144–146] warn against reaching hasty conclusions concerning direct comparisons between mono-, di-, and trialkylamines.

Alkylamines

Here we are concerned mainly with alkylamines. Detailed SCF charge analyses [43] indicate that any gain in total charge translates into a downfield ^{15}N shift, which is the trend exhibited by alkylamines, but also by nitroalkanes and isonitriles. Examples are offered in Table 6.5 (see also Fig. 6.6), along with pertinent ionization potentials, indicated in $kcal/mol^{-1}$.

The comparison with adiabatic ionization potentials [147] indicates that the latter decrease as electronic charge builds up on nitrogen, as one would normally expect, thus suggesting that the charges calculated here are in the right order. (These ionization potentials correspond to the suppression of an electron of the lone pair on

TABLE 6.5. Nitrogen Net Charges and ^{15}N NMR Shifts

Molecule[a]	q_N (me)	$\delta(^{15}N)$[b]	IP (kcal/mol)
1 CH_3NH_2	−374.4	371.1	206.8
2 $iso\text{-}C_4H_9NH_2$	−381.5	356.5	—
3 $n\text{-}C_3H_7NH_2$	−384.2	353.4	202.5
4 $C_2H_5NH_2$	−384.6	349.2	204.3
5 $iso\text{-}C_3H_7NH_2$	−392.2	331.9	201.1
6 $tert\text{-}C_4H_9NH_2$	−397.3	318.1	199.2
7 $(CH_3)_2NH$	−308.5	363.3	190.0
8 $(CH_3)(C_2H_5)NH$	−316.6	345.8	—
9 $(n\text{-}C_3H_7)_2NH$	−323.3	334.3	—
10 $(C_2H_5)_2NH$	−325.4	327.5	184.7
11 $(CH_3)_3N$	−255.1	356.9	180.3
12 $(CH_3)_2(C_2H_5)N$	−268.0	345.1	—
13 $(C_2H_5)_2(CH_3)N$	−280.8	334.3	—

[a]The numbering corresponds to that of Fig. 6.6.
[b]$\delta(^{15}N)$ in ppm from HNO_3, in methanol, from Ref. 149.

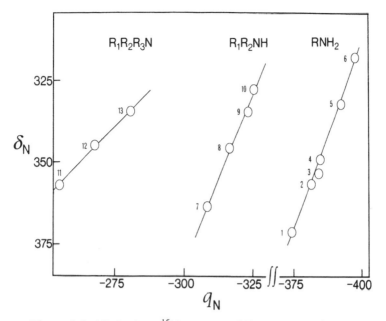

Figure 6.6. Alkylamines: ^{15}N resonance shifts versus net charges.

nitrogen [148].) On the other hand, a gain in electronic charge also appears to come with a concurring downfield ^{15}N chemical shift. [In the convention adopted here for the ^{15}N resonance shifts [149] (ppm from HNO_3, in methanol), increasingly larger $\delta(^{15}$N$)$ values indicate upfield resonance shifts.]

The charges reported in Table 6.5 are STO-3G results. They require a rescaling; that described by $\Delta q_C^{rescaled} \approx (35.1/69.4)\Delta q_C^{STO-3G}$ is a reasonable approximation for paraffins [44] and—tentatively adopted for the alkylamines [139]—proved adequate.

Separate correlation lines are observed for mono-, di-, and trialkyl-amines; namely, after proper rescaling [139], we obtain

$$q_N = 0.218\big(\delta(^{15}N) - 371.1\big) - 9.00 \text{ (me) (primary amines)} \tag{6.12}$$

$$q_N = 0.247\big(\delta(^{15}N) - 363.3\big) - 23.35 \text{ (me) (secondary amines)} \tag{6.13}$$

$$q_N = 0.168\big(\delta(^{15}N) - 356.9\big) - 29.50 \text{ (me) (tertiary amines).} \tag{6.14}$$

Adding to the theoretical difficulties, there is also the problem of assessing the appropriateness of the NMR data selected for use in comparisons like those described here, because solvent and concentration effects are often much larger for nitrogen-15 than for carbon-13 NMR spectra [150]. Litchman and coworkers [151], for example, investigated in detail the effects of binding of the solvent protons to the nitrogen lone pair and (or) influences of solvent lone pairs binding with the protons on nitrogen, concluding that different solvent effects are generally to be expected for primary and tertiary amines, and probably for secondary amines as well. Besides hydrogen bonding, other possible causes for the solvent effects on ^{15}N chemical shifts include polarization of solute molecules in a very polar medium and conformational changes resulting from strong solvent–solute interactions. The importance of these interactions is revealed by the correlations observed for the ^{15}N NMR shifts of the saturated amines with the ^{13}C shifts of the corresponding carbons in analogous hydrocarbons [149,152–154], namely, by the very existence of separate, solvent-sensitive, linear correlations for primary, secondary, and closely related groups of tertiary amines. Results of this sort, obtained with great care for a large variety of compounds and interpreted in detail by Roberts and coworkers [149,152–154], are instrumental in a more balanced assessment of the charge–shift correlations [Eqs. (6.12)–(6.14)]. It appears, indeed, that the charges to be used in this type of correlation should not be those of the isolated amines but rather those of the (still poorly known) aggregates as they are physically present in shift measurements. However, the linearity of the ^{15}N/^{13}C correlations, as well as the corresponding ones between atomic charges, supports the overall validity of correlations involving ^{15}N shifts and nitrogen charges of model isolated amines, at least in terms of general charge–shift trends within series of closely related compounds.

Hence, despite the theoretical and solvent-related difficulties outlined above, it may be concluded that any increase in electronic charge at amine nitrogen atoms translates into a downfield chemical shift.

The (at least approximate) validity of Eqs. (6.12)–(6.14) is clearly demonstrated by the remarkable quality of intrinsic CN bond energies and of bond dissociation energies calculated by means of nitrogen net atomic charges deduced in this manner.

Nitroalkanes

In nitroalkanes, R—NO$_2$, a regular upfield shift of the nitrogen resonance with increasing electronegativity of R is observed and obeys simple additivity rules [155]. Upon replacing hydrogen atoms in CH_3NO_2 with alkyl groups, a downfield shift of $\sim$10 ppm takes place—a trend similar to that encountered with amine nitrogen resonances. This trend visibly reflects the usual electron-releasing ability of alkyl groups, in the order tert-C_4H_9 > iso-C_3H_7 > $\cdots$ > CH_3.

STO-3G charge analyses [43] and the appropriate ^{15}N NMR shifts [156] indicate that a gain in electronic charge at nitrogen translates linearly into a downfield NMR shift. Interestingly, the slope is similar to those found for mono- and dialkylamines [139]:

$$q_N - q_N^{CH_3NO_2} = 0.253 \left(\delta_N - \delta_N^{CH_3NO_2} \right) \tag{6.15}$$

Isonitriles

The nitrogen resonance signal of isonitriles moves to higher fields [157] with increasing electronegativity of the group R in R—NC. The shift is very regular, about 15 ppm downfield for each H atom in CH_3—NC replaced with an alkyl group.

This resembles the trends observed for amines and nitro groups. STO-3G charges [43] show that the nitrogen signal [157] is farther downfield as electronic charge increases at this atom.

Substituted Pyridines and Diazines

Using the experimental geometry of pyridine [158], the STO-3G optimization of its scale factors was performed until charges were stable to within ~0.01 me. Taking advantage of Del Bene's work [159] showing that the pyridine ring is essentially unchanged in the equilibrium structures of 4-R-pyridines, only the geometries and the scale factors of the substituents were subsequently optimized for the 4-substituted derivatives, under the appropriate symmetry constraints (C_{2v} for NO_2 and NH_2, C_s otherwise). The results are indicated in Table 6.6.

Electron-releasing substituents (NH_2, OH, OCH_3, CH_3) appear to increase the π (and total) charge of the ring nitrogen atom, whereas the opposite is true for electron-withdrawing groups (NO_2, CHO, CN, $COCH_3$), in accord with common views. These results match those of Hehre et al. [127], showing that carbon-4 in substituted benzenes gains or loses charge depending on whether the substituent donates or withdraws electrons. Earlier studies by Pople and coworkers [160] on related systems similarly concluded that theoretical SCF charge distributions do, indeed, support many of the ideas of classical organic chemistry.

The parallelism between the behavior of nitrogen in 4-substituted pyridines and that of the corresponding carbon in substituted benzenes is well illustrated in Fig. 6.7. The results of Table 6.6 reveal another important similarity between

TABLE 6.6. Nitrogen σ, π, and Total $\sigma + \pi$ Net Charges (me) and ^{15}N NMR Shifts of 4-Substituted Pyridines and Selected Azines

Molecule[a]		q_σ	q_π	q_{total}	$\delta(^{15}N)$[b]
1	HO-pyr	−175.5	−95.3	−270.8	−138
2	NH_2-pyr	−176.2	−96.6	−272.8	−42
3	OCH_3-pyr	−189.0	−69.9	−258.9	−27
4	CH_3-pyr	−190.6	−69.7	−260.3	−11
5	H-pyr	−195.5	−59.6	−225.1	0
6	CN-pyr	−200.5	−39.7	−240.2	9
7	$COCH_3$-pyr	−199.0	−51.2	−250.2	25
8	NO_2-pyr	−204.4	−28.8	−233.1	28
9	CHO-pyr	−199.4	−49.3	−248.7	29
10	1,4-Diazine	−196.0	−18.8	−214.8	46.1
11	Pyridine	−186.6	−53.8	−240.4	63.5
12	1,3-Diazine	−180.8	−84.1	−264.9	84.5
13	1,3,5-Triazine	−183.3	−101.3	−284.6	98.5

[a]Molecules **1–9** were partially optimized as described in the text; the results for **10–13** were derived with full (geometry and scale factor) optimization.
[b]$\delta(^{15}N)$ in ppm from pyridine for **1–9**; ppm from external $MeNO_2$ for **10–13**.
Source: Ref. 150.

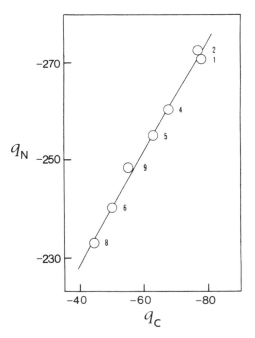

Figure 6.7. Comparison between the nitrogen net charges of selected 4-R-pyridines and those of C-4 in the analogous substituted benzenes. (A similar correlation holds for the corresponding π charges.) The numbering is reported in Table 6.6. (Reproduced with permission from Ref. 43.)

4-pyridine nitrogens and aromatic carbons. As is the case for the latter, σ and π populations of the ring nitrogens vary in inverse manner; the changes in π charge predominate [43]. Finally, the correlation between resonance shifts and atomic charges of the pyridine nitrogen atom reveals an upfield shift accompanying a gain in total $\sigma + \pi$ population, just like their aromatic carbon analogs.

Of course, because of the linear inverse relationship between σ and π charges, the correlation with total charges also implies the existence of individual linear correlations with σ and π populations, namely, upfield shifts for increasing π-electron densities, just as in the case of aromatic hydrocarbons. The latter result also follows from Pariser–Parr–Pople calculations of π charge densities reported by Witanowski et al. [161]. All of these considerations apply equally well to the series including pyridine, 1,3-diazine, 1,4-diazine, and 1,3,5-triazine, as revealed by their ^{15}N spectra in DMSO solution [162] (Fig. 6.8).

Solvent and concentration effects should obviously not be disregarded in the present evaluation of charge–shift correlations. Shielding solvent effect affecting the ^{15}N resonances of pyridines, and (more generally) of sp^2-hybridized nitrogens, are associated with hydrogen bonding to the nitrogen unshared pairs. Such hydrogen bonding modifies the $\pi \rightarrow \pi^*$ transition energies and, hence, the paramagnetic screening that appears to be the dominant influence on the chemical shifts of these

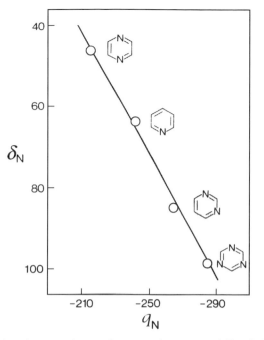

Figure 6.8. Comparison between the net charges and resonance shifts of nitrogen in pyridine, 1,3-diazine, 1,4-diazine, and 1,3,5-triazine (me). (Reproduced with permission from Ref. 43.)

compounds [163]. Because of a careful selection of suitable "standard" experimental conditions, however, these solvent effects should not impair the overall validity of the present conclusions, namely, that any gain in total electronic charge at azine nitrogens translates into an upfield shift—in sharp contrast with the results obtained for amines, nitroalkanes, and isonitriles.

6.9 CORRELATIONS INVOLVING O-17 ATOMS

This section is about dialkylethers, aldehydes, ketones and alcohols. Oxygen net charges deduced from standard STO-3G calculations are indicated in Table 6.7 (see also Fig. 6.9), along with their NMR shifts (ppm from water) for dialkylethers [140] and carbonyl compounds [140], as well as selected ionization potentials [147].

A comparison of the ionization potentials of selected ethers with their oxygen net charges[2] yields a correlation, $IP = 0.0274\,q_o + 18.16$ eV (with an average error of 0.040 eV and a correlation coefficient of 0.9914), and reflects the expectation that electron withdrawal becomes easier as the oxygen atom becomes electron-richer. This result illustrates the ordering of the oxygen charges in agreement with all known aspects related to the inductive effects of alkyl groups. Thus it can be

[2]Optimized STO-3G Mulliken net charges were used.

TABLE 6.7. ^{17}O **NMR Shifts and Net Charges (me) of Oxygen Atoms**

Molecule[a]	q_O (me)	δ_O	IP (eV)[b]
1 $(CH_3)_2O$	−297.1	−52.5	10.00
2 $CH_3OC_2H_5$	−306.0	−22.5	—
3 $CH_3Oiso\text{-}C_3H_7$	−313.3	−2.0	—
4 $CH_3Otert\text{-}C_4H_9$	−316.6	8.5	—
5 $(C_2H_5)_2O$	−314.9	6.5	9.53
6 $C_2H_5Otert\text{-}C_4H_9$	−325.3	40.5	—
7 $(iso\text{-}C_3H_7)_2O$	−330.1	52.5	9.20
8 $iso\text{-}C_3H_7Otert\text{-}C_4H_9$	−333.4	62.5	—
9 $(tert\text{-}C_4H_9)_2O$	−334.1	76.0	—
CH_3CHO	−228.9	592.0	—
C_2H_5CHO	−229.6	579.5	—
$iso\text{-}C_3H_7CHO$	−230.1	574.5	—
$(CH_3)_2CO$	−266.7	569.0	—
$CH_3COC_2H_5$	−269.5	557.5	—
$CH_3COiso\text{-}C_3H7$	−268.1	557.0	—
$(C_2H_5)_2CO$	−272.1	547.0	—
$C_2H_5COiso\text{-}C_3H_7$	−270.7	543.5	—
$(iso\text{-}C_3H_7)_2CO$	−273.8	535.0	—

[a]The numbering corresponds to the points in Fig. 6.9.
[b]For di-n-propylether, it is IP = 9.27 eV.

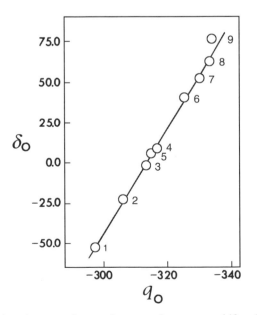

Figure 6.9. Comparison between the net charges and resonance shifts of oxygen in selected dialkylethers. (Reproduced with permission from Ref. 41.)

assumed with confidence that the oxygen atom in, say, di-*tert*-butylether is electron-richer than that of dimethylether.

The results indicate that any gain in electronic charge at the oxygen atom of dialkylethers is accompanied by a downfield ^{17}O NMR shift.

It is clear that the trends in electron populations at the ROR' ether oxygen atoms reflect the electron-releasing (or withdrawing) abilities of the R,R' alkyl groups, just as is the case with the methylene carbons in RCH_2R' hydrocarbons. Since for the ether oxygen and the sp^3 carbons it now appears that chemical shifts and atomic charges are linearly related to one another, it follows that the ^{17}O NMR shifts of the ROR' ethers are expected to correlate with the methylene ^{13}C shifts of the corresponding RCH_2R' hydrocarbons. This is, indeed, the case, as demonstrated convincingly by Delseth and Kintzinger [140]. Moreover, similar correlations [140] between the ^{13}C shifts of the carbon atoms in ROR' ethers and those of the "parent" RCH_2R' hydrocarbons clearly reflect the close correspondence in the structure-related effects that govern electron distributions indicating, namely, that the individual C atoms of the alkyl part of the ethers behave quite like hydrocarbon C atoms.

Equation (6.9) describes the shift–charge correlation for paraffins. In order to derive an expression for ^{17}O nuclei on the same footing as (6.9), an extensive geometry and exponent optimization was carried out for the dimethyl-, diethyl-, and diisopropylethers. Under these conditions, the oxygen STO-3G Mulliken charges of dimethylether (-267 me), diethylether (-295 me), and diisopropylether (-322 me) indicate that the δ_O/q_O slope is about $(1/1.8)$th that of the corresponding δ_C/q_C slope, obtained from the same basis set, i.e.

$$\frac{\Delta\delta_O}{\Delta q_O} \simeq \frac{\Delta\delta_C/\Delta q_C}{1.8}$$

Consequently, we deduce from Eq. (6.9) that

$$q_O - q_O^{(C_2H_5)_2O} = -0.267\left(\delta_O - \delta_O^{(C_2H_5)_2O}\right) \text{ (me)} \qquad (6.16)$$

where $q_O^{(C_2H_5)_2O} \simeq 5.18$ me [44], as revealed by energy calculations, with the diethylether ^{17}O shift at 6.5 ppm from water.

Finally, regarding the carbonyl oxygen atoms, the lack of precision (1–2 me) accompanying standard STO-3G calculations and, above all, the narrow range of Δq_O variations, makes it difficult to assess how accurately carbonyl-^{17}O NMR shifts and atomic charges are related to one another. Energy analyses [141] suggest the following formula

$$\Delta q_O \simeq 2.7\left(\delta_O - \delta_O^{acetone}\right) \qquad (6.17)$$

which is tentative and could well be subject to future revisions. (Note, however, that the calculated and experimental atomization energies of carbonyl compounds thus far agree within 0.20 kcal/mol (root-mean-square deviation).

Finally, energy analyses of a collection of alcohols [35] suggest the following formula

$$q_O - q_O^{H_2O} \simeq -0.165 \, \delta_O \qquad (6.18)$$

where δ_O is the ^{17}O shift relative to external water. Again, this must be considered as a tentative proposal, advocated by its applications in energy calculations.

6.10   SUMMARY

In series of closely related compounds, the change in total (dia+para)magnetic shielding of atomic nuclei is nearly that of the *local* paramagnetic term, because of important cancellation effects involving nonlocal dia- and paramagnetic contributions. This offers a justification for relationships between nuclear magnetic resonance shifts and local atomic populations that are observed. It is important, however, to consider the type ($2s$, $2p$, σ, or π) of electrons that are responsible for the variations in atomic charges. Correlations between ^{13}C NMR shifts and atomic populations of aromatic compounds, for example, should not be interpreted in terms of π electrons only, because the slope of shift versus π charge (i.e., the ~ 160 ppm/electron value that is usually invoked) does not describe an intrinsic effect of π charges on magnetic shielding but accounts for the fact that σ and π charges vary in opposite directions in this class of compounds. The explicit consideration of the inverse variations of σ and π charges, where appropriate, offers an explanation for the observation that charge–shift correlations can have positive or negative slopes. It appears, indeed, that an increase in total electronic population is accompanied by (1) a high-field shift when the electron enrichment results from a gain in π charge prevailing over the concurrent loss in σ electrons (aromatic and olefinic C, carbonyl O atoms, and sp^2 N atoms) or (2) a downfield shift when the increase in charge is dictated by that of the σ population (sp^3 C, dialkyl ether O atoms, and nitrogen of amines, for example).

So, on the face of things, correlations of NMR shifts with atomic charges look like a settled argument. Yet, because of the LCAO-MO Mulliken charges which were used, it stands to reason that the numbers should not be taken as they come. But thanks to the intrinsic qualities of these SCF analyses, all of which systematically and not unexpectedly reproduce the same general trends familiar to chemists [27,38,96], the essence of charge–shift correlations is not at stake; it is the numbers that require attention. An important clue, embodied in Eq. (6.7), is offered by the alkanes; supporting evidence include CI charge analyses (described in Chapter 5) and extensive verifications by means of energy calculations (Chapter 13). The latter now play a major role in the appropriate definition of other charges, such as those of oxygen and nitrogen, fit for correlations with NMR shifts.

Correlations between the shifts of atoms in analogous positions highlight the persistent role of the inductive effects. So, for example, the nitrogen charges of

4-*R*-pyridines correlate with those of C-4 in the analogous substituted benzenes (Fig. 6.7); similarly, the ^{17}O shifts of ethers correlate with the methylene ^{13}C shifts of the corresponding hydrocarbons. Roberts et al. [164] have shown that a simple linear correlation exists between the chemical shifts of carbon atoms in alcohols and the corresponding hydrocarbon wherein a methyl group replaces the oxygen function. In this vein, Eggert and Djerassi [165] showed that a similar correlation exists for amines; as a consequence, it is possible to predict the chemical shifts of amines, since the shifts of the C atoms of alkanes can be calculated using the parameter set of Lindeman and Adams [166].

Of course, this list is by no means exhaustive. Additional correlations of this nature would not only broaden our views in the field of charge–shift relationships but also enhance our ability of predicting bond and molecular energies with chemical accuracy by means of charges deduced from NMR shifts.

Therein lies the merit of charge–shift correlations.
We could take advantage of them.
Ecce tempus: nunc aut nunquam!

CHAPTER 7

CHARGES AND IONIZATION POTENTIALS

This inroad into the study of the adiabatic ionization potentials (IPs) of paraffins and a hypothetical correlation with atomic charges is to determine (1) whether such a correlation exists in a first place and (2) if so, what sort of charges would satisfy it.

A crude justification for such a study can be found in the work of Widing and Levitt [167], who described numerous correlations between adiabatic ionization potentials and Taft's inductive substituent constants (an indirect way of correlating IPs with charges) and in a note by Streitwieser [168], who justified to some extent a dependence of IPs on local charge densities.

Here attention focuses primarily on the charges to be used in such a correlation. Rather than involving SCF atomic charges, advantage is taken from the general formulation described earlier (Chapter 5), using a set defined by letting $q_C^{C_2H_6} = 1$ arbitrary unit for the ethane carbon atom, specifically, a set defined in the so-called relative (dimensionless) scale. These charges are given in Table 7.1 in their general form, with $(3n + x)/3n$ (for primary C atoms), $(2n + x)/3n$ (secondary C), $(n + x)/3n$ (tertiary C), and $-x/n$ (quaternary C atom); the numerical values of x were verified by means of the q_C/q_C° charges predicted by Eq. (6.7) and the ^{13}C shifts determined by Grant and Paul [169]. Thus we have the appropriate expressions for the carbon atomic charges required for the comparison with the adiabatic ionization potentials given in Ref. 147. The formulas correspond to the atoms identified as $\ddot{C}$.

Two issues now need be addressed: (1) which atom(s) of each molecule should be considered for the correlation with the IPs and (2) what is the "proper" n value associated with the equations of Table 7.1. The form of the correlation (linear or other) is

Atomic Charges, Bond Properties, and Molecular Energies, by Sándor Fliszár
Copyright © 2009 John Wiley & Sons, Inc.

TABLE 7.1. Carbon Atomic Charges and Ionization Potentials

Molecule[a]	$q(\text{C-1})$	$q(\text{C-2})$	IP (eV)
CH$_4$	$4(n + 1)/3n$	—	12.98
1 $\ddot{\text{C}}$H$_3\ddot{\text{C}}$H$_3$	1.000	1.000	11.65
2 $\ddot{\text{C}}$H$_3\ddot{\text{C}}$H$_2$CH$_3$	$(3n + 0.55)/3n$	$(2n - 3.8)/3n$	11.07
3 CH$_3\ddot{\text{C}}$H$_2\ddot{\text{C}}$H$_2$CH$_3$	$(2n - 3.35)/3n$	$(2n - 3.35)/3n$	10.63
4 $\ddot{\text{C}}$H$_3\ddot{\text{C}}$H(CH$_3$)$_2$	$(3n + 1.03)/3n$	$(n - 7.7)/3n$	10.57
5 CH$_3\ddot{\text{C}}$H$_2\ddot{\text{C}}$H$_2$CH$_2$CH$_3$	$(2n - 3.46)/3n$	$(2n - 2.8)/3n$	10.35
6 $\ddot{\text{C}}$H$_3\ddot{\text{C}}$(CH$_3$)$_3$	$(n + 0.49)/n$	$-4/n$	10.35
7 CH$_3\ddot{\text{C}}$H$_2\ddot{\text{C}}$(CH$_3$)$_2$	$(2n - 2.95)/3n$	$(n - 7.46)/3n$	10.32
8 CH$_3$CH$_2\ddot{\text{C}}$H$_2\ddot{\text{C}}$H$_2$CH$_2$CH$_3$	$(2n - 2.94)/3n$	$(2n - 2.94)/3n$	10.18
9 (CH$_3$)$_2\ddot{\text{C}}$H$\ddot{\text{C}}$H$_2$CH$_2$CH$_3$	$(n - 7.58)/3n$	$(2n - 2.39)/3n$	10.12
10 CH$_3$CH$_2\ddot{\text{C}}$H(CH$_3$)CH$_2$CH$_3$	$(2n - 3.09)/3n$	$(n - 7.08)/3n$	10.08
11 (CH$_3$)$_3\ddot{\text{C}}\ddot{\text{C}}H_2CH_3$	$-11.85/3n$	$(2n - 2.67)/3n$	10.06
12 (CH$_3$)$_2\ddot{\text{C}}$H$\ddot{\text{C}}$H(CH$_3$)$_2$	$(n - 7.23)/3n$	$(n - 7.23)/3n$	10.02

[a]The numbering corresponds to the points in Fig. 7.1.

not of any particular interest at this stage; all that is required is that the dependence be monotonic.

Several attempts were made, including attempts to correlate the IPs with the net charge of the electron-richest C atom of each molecule or with a combination of the electron-richest carbon–hydrogen pair of each molecule, all of which failed in giving any monotonic dependence of the IPs on these quantities. The only solution is that resulting from comparisons of ionization potentials with the electron-richest pair of bonded C atoms in each molecule. This correlation turns out to be linear (Fig. 7.1), although such a linearity had not been postulated a priori. The least-square

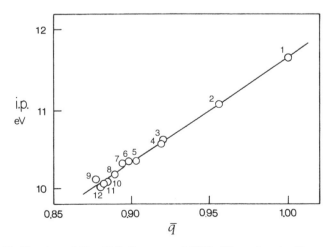

Figure 7.1. Verification of Eq. (7.1) for $n = -4.4083$. (Reproduced with permission from Ref. 170.)

analysis of the data, using the equations of Table 7.1, indicates that

$$n = -4.4083$$
$$\text{IP} = 13.15\,\bar{q} - 1.50 \text{ eV} \tag{7.1}$$

with a standard deviation of 0.040 eV [170].

In this equation, $\bar{q} = \frac{1}{2}[q(\ddot{C}_1) + q(\ddot{C}_2)]$ is one-half the sum of the charges of the electron-richest bonded $\ddot{C}$ atoms in each alkane molecule, expressed by means of the appropriate formulas given in Table 7.1. The net atomic charges q_C and q_C° are both positive because $n < 0$. A decreasing ratio q_C/q_C° thus means that a carbon atom gains electronic charge with respect to that of ethane. Figure 7.1 indicates that the IPs of the alkanes are lowered as the joint electron population of the electron-richest pair of atoms increases.

This result is similar to that obtained by Widing and Levitt for the normal alkanes and to that observed for alkyl-substituted ethylenes [171]; in the latter case, linear correlations were obtained between the IPs and the sum of charges of the unsaturated carbon atoms, whereby any increase of their electron density due to substituent effects leads to a lowering of the molecular ionization potential.

7.1 CONCLUSION

It must now be made clear that in performing this least-square fitting of the ionization potentials (also considering possible contributions of quadratic charge terms), n was allowed to assume freely any value that would give the best result. In this respect, $n = -4.4083$ can be regarded as a truly independent "experimental" n value: it is virtually the value determined from correlations with ^{13}C NMR shifts. This result is certainly significant.

The obvious advantage of the present analysis derives from the use of the charges indicated in Table 7.1, which are equivalent to those given by most theoretical methods—or, should we say, by practically any method—namely, by the familiar ab initio techniques using Mulliken's population analysis, except for the characteristic value of n, which differs from method to method. In that, the most general approach was used for studying a property–charge relationship by letting n be a quantity to be calibrated by experiment without being bothered by specifics in the calculation of atomic charges, such as those pertaining to the pertinence of Mulliken's population analysis, or the selection of basis sets, including the aleas following from basis set superposition errors.

Let us now adopt the more conventional perspective of theoretical charge analyses.

CHAPTER 8

POPULATION ANALYSIS

Knowledge of the molecular wavefunction enables us to determine the electron density at any given point in space. Here we inquire about the amount of electronic charge that can be associated in a meaningful way with each individual atom of a N-electron system. Our analysis covers Mulliken's celebrated population analysis [31], as well as a similar, closely related method.

8.1 THE STANDARD MULLIKEN FORMULA

Mulliken's population analysis is rooted in the LCAO (linear combination of atomic orbitals) formulation; it is not directly applicable to other types of wavefunctions. With $c_{r_k i}$ representing the coefficient of the rth type of atomic orbital ($1s$, $2s$, etc.) of atom k in the ith molecular orbital, we describe the latter by

$$\phi_i = \sum_{r_k} c_{r_k i} \chi_{r_k} \tag{8.1}$$

where the summation extends over all the appropriate normalized basis functions χ_{r_k} and the subindex k labels the different nuclei in the system. The corresponding overlap population associated with atoms k and l due to atomic orbitals of type r and s, respectively, is then

$$2n_i c_{r_k i} c_{s_l i} S_{r_k s_l} \tag{8.2}$$

with $S_{r_k s_l} = \langle \chi_{r_k} | \chi_{s_l} \rangle$, where n_i is the occupation number of that MO. Finally, Mulliken's analysis yields the population N_k on atom k from the appropriate sums over all doubly occupied molecular orbitals i and over all types of basis functions:

$$N_k = \sum_i \sum_r n_i \left(c_{r_k i}^2 + \sum_{l \neq k} c_{r_k i} c_{s_l i} S_{r_k s_l} \right) \tag{8.3}$$

The problem with this analysis is that the selection of one or another basis set dramatically affects the calculated charges and occasionally leads to unphysical results [44]. Overlap populations (8.2) are largely responsible for this situation.

Mulliken's formula for N_k implies the half-and-half (50/50) partitioning of all overlap populations among the centers $k, l, \ldots$ involved. On one hand, this distribution is perhaps arbitrary, which invites alternative modes of handling overlap populations. On the other hand, Mayer's analysis [172,173] vindicates Mulliken's procedure. So we may suggest a nuance in the interpretation [44]: departures from the usual halving of overlap terms could be regarded as ad hoc corrections for an imbalance of the basis sets used for different atoms. But one way or another, the outcome is the same. It is clear that the partitioning problem should not be discussed without explicit reference to the bases that are used in the LCAO expansions.

A modified partitioning procedure is considered in the next section.

8.2 MODIFIED POPULATION ANALYSIS

The population N_k on atom k is now defined as follows [21,44,108]

$$N_k = \sum_i \sum_r n_i \left(c_{r_k i}^2 + \sum_{l \neq k} c_{r_k i} c_{s_l i} S_{r_k s_l} \lambda_{r_k s_l} \right) \tag{8.4}$$

where the weighting factor $\lambda_{r_k s_l}$ causes the departure from the usual halving of the overlap terms. Mulliken's charges correspond to $\lambda_{r_k s_l} = 1$. In terms of the difference

$$\sum_i \sum_r \sum_{l \neq k} n_i c_{r_k i} c_{s_l i} S_{r_k s_l} (1 - \lambda_{r_k s_l}) = \sum_{l \neq k} p_{kl} \tag{8.5}$$

between Mulliken charges [Eq. (8.3)] and those given by Eq. (8.4), one obtains for the net atomic charge, $q_k = Z_k - N_k$, of atom k that

$$q_k = q_k^{\text{Mull}} + \sum_{l \neq k} p_{kl} \tag{8.6}$$

where Z_k is the nuclear charge of atom k and N_k its electron population. We know of no general recipe permitting the calculation of $\sum_{l \neq k} p_{kl}$ but can benefit from the fact that SCF charge analyses given by Eq. (8.4) always reproduce the familiar inductive effects, no matter what $\lambda_{r_k s_l}$ is used [44].

8.3   AN ADEQUATE APPROXIMATION

A simple strategy has been developed for saturated hydrocarbons [44,96]. Assume $p = 0$ for any carbon–carbon bond and $p = $ constant for any carbon–hydrogen bond. Hence

$$q_C = q_C^{Mull} + N_{CH} \times p \tag{8.7}$$

$$q_H = q_H^{Mull} - p \tag{8.8}$$

where N_{CH} represents the number of hydrogen atoms attached to any given carbon. The idea is simple; if $p = p_{CH} = -p_{HC}$ is the error made by assuming Mulliken's half-and-half partitioning of any CH overlap population (i.e., $\lambda = 1$), the error made in the computation of the net charge on carbon depends on the number of hydrogens attached to it. In other words, we come back to Eqs. (5.10) and (5.11), but with an interpretation attached to them.

These corrections mean that n changes as well [Eq. (5.14)]. Hence, after the correction of an original set of Mulliken charges to ensure the proper scaling represented by n, a second correction is in order to ensure the proper "absolute" values of these charges. Finally, we shall use the following corrected carbon and hydrogen net charges, based on a set of Mulliken results, in lieu of the original ones indicated in Eqs. (8.7) and (8.8):

$$q_C = \Lambda(q_C^{Mull} + N_{CH}p) \tag{8.9}$$

$$q_H = \Lambda(q_H^{Mull} - p) \tag{8.10}$$

In this form, it becomes possible to analyze the merits of Mulliken charge distributions in comparisons with physical observables. Namely, we want to learn the "true" value of n and the appropriate value of Λ for given choices of basis sets.

Three approaches were followed to this end:

- Comparison of adiabatic ionization potentials (IP) of normal and branched alkanes with carbon net charges, which indicates a lowering of the IPs with increasing electron population of the electron-richest bonded pair of carbon atoms in the molecule [170]. A monotonic correlation (which turns out to be linear and remarkably accurate) is possible *only* with atomic charges adjusted for $n = -4.4083$ and the corresponding p given by Eq. (5.14).
- Comparison of carbon-13 NMR shifts with charges of saturated carbons, as defined in Eq. (8.9), showing that a highly accurate linear correlation exists, but *only* with charges adjusted for $n = -4.4122$, corresponding to $p = 30.12$ me for the fully optimized STO-3G Mulliken charges that were used [38,39]. Practically the same p correction applies to ethylenic carbon atoms calculated with the same STO-3G basis, in comparisons with their ^{13}C NMR shifts [40].
- Direct calculations of atomization energies, ΔE_a^* (see Chapter 13), using formulas (8.9) and (8.10) with Mulliken charges obtained from the fully optimized

STO-3G basis and experimental results for ΔE_a^*. A least-square regression led to $p = 30.3 \pm 0.3$ me and $n = -4.446 \pm 0.057$ [108]. Concurrently it became possible to estimate the value of the reference charge, that of the ethane carbon atom: 35.1 me. Similar work for ethylenic molecules led to q_C (C_2H_4) = 7.7 me. For comparison, calculations using a carefully optimized 4-31G basis and configuration interaction gave 37.8 me for the carbon net charge of ethane, and 7.5 me for that of ethylene [51]. In short, $\Lambda \to 1$ with CI wavefunctions.

The nature of the charges to be used in our bond energy formula is thus unmistakably identified. Our best estimate is probably that offered by the most accurate NMR chemical shift correlations, $n = -4.4122$. This result is attractive for its physical content; any set of Mulliken charges (i.e., irrespective of what basis set is used), corrected with the p value that produces carbon charges corresponding to $n \approx -4.41$, does in fact end up with carbon atomic charges that are as similar as possible to one another. "Almost nothing" really seems to happen with the alkyl carbon and hydrogen atoms in going from one molecule to another—Nature resists changes—but that little bit is precisely what matters.

The modified population analysis [Eq. (8.4)] appears to be adequate. The correction suggested by Fig. 5.2 is now readily understood in terms of Eqs. (8.4)–(8.6) as a basis set effect.

The arguments presented so far can be extended with reasonable confidence to other bonds, such as carbon–nitrogen bonds, where an original net charge on k, q_k, is rescaled as follows to give q_k^{rescaled}:

$$q_k^{\text{rescaled}} = q_k - \sum_{l \neq k} p_{kl} \tag{8.11}$$

This is as far as we get with Mulliken-type methods.

Let us briefly comment on the $X\alpha$ scattered-wave (SW) method [174,175] which does not involve Mulliken's population analysis.

In contrast to ab initio methods, SCF–$X\alpha$–SW theory approximates the exchange potential by a local exchange potential [175] that greatly simplifies the computations of many-electron systems. Furthermore, the one-electron (nuclear attraction + Coulomb + exchange) potential is treated by the "muffin-tin approximation." Briefly, this means that each atom in the molecule is surrounded by a sphere, and that in each such atomic sphere the exact one-electron potential is replaced by a spherical average. The atomic sphere regions are designated as regions I. The region between the spheres is the so-called intersphere region II where the potential is assumed to be constant. Finally, the whole molecule is surrounded by a sphere, the extramolecular region III, where the potential is replaced by a spherical average. One of the main advantages of the SW model is that the MO wavefunctions are described as rapidly converging multicenter partial wave expansions, whose radial parts are given by numerical integration. Hence the method is independent of basis set problems, avoiding thus the difficulties encountered in the LCAO approach with the choice of the proper basis functions. But things are not that simple. The charge in each atomic region is considered as the net atomic population, and the charge in region III is distributed among only those atoms that touch the outer sphere. In the intersphere region, however, there is no distinctive way of assigning charges, which results in some arbitrariness in their partitioning and, hence, in describing charge distributions. (For example, one particular partitioning technique [176] was implemented [104] with some modifications related to its more detailed utilization to get the results of Table 5.4.) Thus, despite the inherent merits of the method, it is unfortunately of little help in producing reliable charge distributions.

8.4 CONCLUSIONS

We have learned about the unique ordering of the carbon net charges relative to one another. All methods using Mulliken's population analysis, both ab initio and semiempirical, no matter what basis sets are used to construct the wavefunctions, reproduce the following sequence of inductive effects:

$$CH_3 < CH_3CH_2 < (CH_3)_2CH < \cdots < (CH_3)_3C$$

The converse message is clear and devoid of circumlocution; all methods that reproduce the inductive order are equivalent—they all give the same answer after an appropriate rescaling of overlap terms [27].

This rescaling reflects the idea that any increase of electronic charge at a center, as a consequence of an enrichment of the basis functions describing it, is unphysical if, lacking equipoise, atoms bonded to it suffer from poorer basis set descriptions. The parameter $\lambda_{r_k s_l}$ introduced in Eq. (8.4) is there to correct this imbalance if we follow Mayer's claim [172,173] that Mulliken's half-and-half partitioning of overlap terms between the concerned atoms should not be tampered with. It is felt that the way $\lambda_{r_k s_l}$ depends on the basis sets used for describing atoms k and l deserves attention as part of an effort aimed at letting $\lambda_{r_k s_l}$ approach Mulliken's limit $\lambda = 1$ as closely as possible.

This observation offers a way of reconciling such disparate sets of charge results as they are obtained either from different basis sets in SCF ab initio calculations or else in comparisons involving semiempirical results. Equation (8.11) represents an adequate description.

PART II

CHEMICAL BONDS: ENERGY CALCULATIONS

CHAPTER 9

THERMOCHEMICAL FORMULAS

This chapter offers pertinent information facilitating the comparison between theory and experiment using thermochemical data, such as the standard enthalpies of formation, and spectroscopic information that serve our purpose. Special attention is given to zero-point and heat-content energies which are often not as readily available as desired.

9.1 BASIC FORMULAS

The atomization

$$\text{Molecule} \longrightarrow n_1 A_1 + n_2 A_2 + \cdots + n_k A_k$$

of a given molecule into its constituent n_1 atoms A_1, n_2 atoms A_2, and so on provides a measure for chemical binding. The relevant thermochemical information is usually expressed in terms of enthalpy of formation ΔH_f or enthalpy of atomization ΔH_a of the molecule under consideration

$$\Delta H_a = \sum_k n_k \Delta H_f(A_k) - \Delta H_f \qquad (9.1)$$

Atomic Charges, Bond Properties, and Molecular Energies, by Sándor Fliszár
Copyright © 2009 John Wiley & Sons, Inc.

where the $\Delta H_f(A_k)$ terms are the enthalpies of formation of the gaseous atoms A_k. But for an isolated molecule we refer more appropriately to the *energy* of atomization, ΔE_a:

$$\Delta E_a = \Delta H_a - \left(\sum_k n_k - 1\right) RT \tag{9.2}$$

The values ΔE_a and ΔH_a are considered at some temperature, usually 25°C, specifically, under working conditions of practical interest. These quantities include contributions from internal rotations that are more or less free in some cases and hindered in others, but need not be considered at 0 K. Zero-point vibrational energies are to be taken into proper account, as these energies, like the thermal ones, cannot be fairly apportioned among bonds or atoms in a molecule since they are not truly additive properties, nor can they be regarded as a part of chemical binding. These reasons prompt us to study a molecule in its hypothetical vibrationless state at 0 K, whose atomization energy is ΔE_a^*.

The relationship between ΔE_a^* and ΔE_a is given by Eq. (9.3)

$$\Delta E_a = \Delta E_a^* - \sum_i f(v_i, T) + \frac{3}{2}\left(\sum_k n_k - 2\right) RT \tag{9.3}$$

which states that the energy of atomization at, say, 25°C is that of the hypothetical vibrationless molecule at 0 K *less* the sum $\sum_i f(v_i, T)$ (over $3\sum_k n_k - 6$ degrees of freedom) of vibrational energy corresponding to the fundamental frequencies v_i, which is already present in the molecule at 25°C. The term $3\left(\sum_k n_k - 2\right)RT/2$ now accounts for the formation of $\sum_k n_k$ atoms with translational energy and the disappearance of one nonlinear molecule with three translational and three rotational contributions of $RT/2$ each. It follows from Eqs. (9.1)–(9.3) that

$$\Delta E_a^* = \sum_k n_k\left[\Delta H_f^\circ(A_k) - \frac{5}{2}RT\right] + \sum_i f(v_i, T) + 4RT - \Delta H_f^\circ \tag{9.4}$$

where all enthalpies are now referred to standard conditions (gas, 298.15 K). The vibrational energy may be separated into a zero-point energy (ZPE) term and a thermal vibrational energy term E_{therm}. Hence

$$\sum_i f(v_i, T) + 4RT = ZPE + E_{therm} + 4RT \tag{9.5}$$

On the other hand, $E_{therm} + 4RT$ is the increase in enthalpy $(H_T - H_0)$ of nonlinear molecules due to their warming up from $T = 0$ to $T = T$. Consequently, Eq. (9.4) can now be written as follows:

$$\Delta E_a^* = \sum_k n_k\left[\Delta H_f^\circ(A_k) - \frac{5}{2}RT\right] + ZPE + (H_T - H_0) - \Delta H_f^\circ \tag{9.6}$$

The same formula applies to linear molecules, except for $H_T - H_0$, which is now $E_{\text{therm}} + \frac{7}{2}RT$.

Equation (9.6) is now the basic formula permitting the comparison between calculated ΔE_a^* results and thermochemical information. The appropriate standard enthalpies of formation of the atoms, $\Delta H_f^\circ(A_k)$, are [177] ΔH_f° (C) $= 170.89$, ΔH_f°(H) $= 52.09$, ΔH_f°(N) $= 113.0$, and ΔH_f°(O) $= 59.54$ kcal/mol (gas, 298.15 K).

9.2 ZERO-POINT AND HEAT CONTENT ENERGIES

Zero-point energies are obtained from vibrational spectra using experimental frequencies whenever available, while the inactive frequencies are extracted from data calculated by means of an appropriate force-field model. In the harmonic oscillator approximation, the zero-point energy is

$$\text{ZPE} = \frac{1}{2} \sum_i h\nu_i \tag{9.7}$$

and the thermal vibrational energy is given by Einstein's formula

$$E_{\text{therm}} = \sum_i \frac{h\nu_i}{\exp{(h\nu_i/kT)} - 1} \tag{9.8}$$

from which $H_T - H_0 = E_{\text{therm}} + 4RT$ is readily deduced.

The infrared and Raman fundamental frequencies ν_i are usually expressed as wavenumbers, in cm^{-1} units (reciprocal centimeters). Taking Planck's constant at $h = 6.6256 \times 10^{-27}$ erg $\cdot$ s, Boltzmann's constant at $k = 1.38054 \times 10^{-16}$ erg $\cdot$ K^{-1} and $c = 2.997925 \times 10^{10}$ cm/s for the speed of light, it is for wavenumbers ω_i expressed in cm^{-1}:

$$\frac{1}{2}h\nu_i = 1.42956 \times 10^{-3}\omega_i \quad \text{kcal mol}^{-1}$$

$$\frac{h\nu_i}{kT} = 4.82572 \times 10^{-3}\omega_i \quad \text{at 298.15 K}$$

The zero-point plus heat content energies can be calculated from the complete set of fundamental frequencies. Everything seems fine, except for the fact that resolved complete vibrational spectra are seldom as readily available as desired, thus rendering bona fide comparisons between theory en experiment difficult. Hence our interest in alternate rules permitting the construction of reliable vibrational energies in a simple manner. It is fortunate that ZPE $+ H_T - H_0$ energies obey, to a good approximation, a number of additivity rules [27,44]. A theoretical foundation for this additivity, which has a long history [178], has been established [27,179]. Useful formulas follow.

Alkanes

A detailed study [180] of acyclic alkanes has revealed that the following formula, in kcal/mol, is accurate

$$ZPE + (H_T - H_0) = 11.479 + 18.213 n_C - 0.343 n_{br} \qquad (9.9)$$

where n_C is the number of carbon atoms and n_{br} the number of branchings. The average deviation between the results deduced in this manner and their spectroscopic counterparts is 0.125 kcal/mol (~ 80 cm^{-1}). The change in $ZPE + (H_T - H_0)$ associated with methyl substitutions giving quaternary carbon atoms, however, can be estimated only as a rough average, $\sim 17.55 \pm 0.25$ kcal/mol.

Cycloalkanes constructed from six-membered rings behave in essence like acyclic alkanes, provided that the suppression of internal rotations is adequately taken into account [180] by subtracting $RT/2 = 0.296$ kcal/mol (at 298.15 K) for each CC bond in the cycle. Most detailed calculations have dealt with six-membered cyclo-alkanes [36,181].

They were conducted at two levels of theory using both a conventional uncorre-lated ab initio Hartree–Fock procedure with a 6-31G(d) basis and a density functional approach. The HF/6-31G(d) results could have served the purpose [182], as, indeed, it turned out in retrospective, but it was decided from the outset to go as far as possible in the effort to catch any nuance that could differentiate one situation from another. The advent of density functional theory (DFT) provides an alternative means of including electron correlation. A method of choice in the study of vibrational frequencies is the hybrid B3LYP [183] procedure that uses Becke's three-parameter exchange functional (B3) [184,185] coupled with the correlation functional of Lee, Yang, and Parr (LYP) [20], in conjunction with the standard 6-311G(d,p) basis. Calculations are best carried out with the GAUSSIAN 94/DFT package of ab initio programs [186]. They include geometry optimizations. The appropriate scaling factors for estimating fundamental frequencies from theoretical harmonic frequencies are readily determined on the basis of the results given for cyclohexane and methylcyclohexane [187]. They are 0.90343 for the HF/6-31G(d) theoretical frequencies and 0.9725 for those obtained from B3LYP/6-311G(d,p) calculations. The scaled ZPE and $H_T - H_0$ results are reported in Table 9.1 for the HF/6-31G(d) set. (There is no significant difference between the present scaled DFT and HF results, particularly in regard to the sum $ZPE + H_T - H_0$.) The present $H_T - H_0$ results do agree with the experimental ones [188], with a root-mean-square deviation of 0.06 kcal/mol in the HF/6-31G(d) set.

The important result is that the comparison of isomerides differing in the number of *gauche* interactions (e.g., **6** vs. **7**, **8** vs. **9**, or **10** vs. **11**) reveals no dependence on the number of *gauche* interactions.

In other words, the $ZPE + H_T - H_0$ energies have no saying in the interpretation of *gauche* effects. Another important point is that the theoretical $ZPE + H_T - H_0$

TABLE 9.1. Theoretical Zero-Point and Heat Content Energies of Cycloalkanes, at 298.15 K (kcal/mol)

	Molecule	HF/6-31G(d)	
		ZPE	$H_T - H_0$
1	Cyclohexane	103.42	4.16
2	Methylcyclohexane[a]	120.36	5.19
3	1,1-Dimethylcyclohexane	137.39	6.01
4	*trans*-1,2-Dimethylcyclohexane	137.40	6.10
5	*cis*-1,2-Dimethylcyclohexane	137.61	6.05
6	*cis*-1,3-Dimethylcyclohexane	137.26	6.12
7	*trans*-1,3-Dimethylcyclohexane	137.47	6.09
8	*trans*-1,4-Dimethylcyclohexane	137.28	6.13
9	*cis*-1,4-Dimethylcyclohexane	137.47	6.08
10	1-*cis*-3-*cis*-5-Trimethylcyclohexane	154.19	7.07
11	1-*cis*-3-*trans*-5-Trimethylcyclohexane	154.39	7.03
12	Ethylcyclohexane	137.70	6.08
13	*n*-Propylcyclohexane	154.98	6.97
14	*n*-Butylcyclohexane	172.25	7.84
15	Bicyclo[2.2.2]octane	125.18	4.98
16	*trans*-Decalin	160.45	6.35
17	*cis*-Decalin	160.74	6.28
18	Adamantane	148.15	5.15
19	Twistane	148.19	5.20
20	*trans*–*syn*–*trans*-Perhydroanthracene	217.37	8.51
21	*trans*–*anti*–*trans*-Perhydroanthracene	216.52	9.06
22	1-*trans*-2-*cis*-3-Trimethylcyclohexane	154.46	7.04
23	Isopropylcyclohexane	154.85	6.95
24	1-Methyl-4-isopropylcyclohexane	171.75	7.90
25	*n*-Pentylcyclohexane	189.53	8.72
26	*n*-Hexylcyclohexane	206.81	9.60
27	*n*-Heptylcyclohexane	224.09	10.48
28	*n*-Octylcyclohexane	241.37	11.36
29	*n*-Decylcyclohexane	275.92	13.12
30	*n*-Dodecylcyclohexane	310.48	14.88

[a]Nearly the same result is obtained for the axial form, namely, ZPE = 120.53 and $H_T - H_0$ = 5.18 kcal/mol.

results satisfy the most useful approximation:

$$\text{ZPE} + H_T - H_0 = 11.850(1 - m) + 18.249n - 0.322n_{\text{tert}} - (n_{\text{cycle}} + m - 1)\frac{RT}{2}$$

(9.10)

where n is the total number of carbon atoms and m is the number of cycles; n_{tert} is the number of tertiary carbon atoms and n_{cycle} the number of carbons in the cycle(s).

It should also be noted that $ZPE + H_T - H_0$ increases by only ~ 18.16 kcal/mol for each CH_2 group inserted in the linear alkyl chain carried by a cyclohexyl ring.

The comparison of the genuine theoretical results with those predicted by this approximation shows a root-mean-square (rms) deviation of ~ 0.2 kcal/mol with those obtained in the HF/6-31G(d) calculations reported in Table 9.1. This result is all the more remarkable as it includes polycyclic molecules (**15–21**), boat-cyclohexane structures (**15, 21**), as well as a twist–boat structure (**19**, twistane = tricyclo [4.4.00^{3,8}]decane). The use of this approximation for $ZPE + H_T - H_0$ in problems of thermochemistry is certainly justified. It is noted that Eq. (9.10) occasionally predicts results that are closer to experiment than B3LYP or HF results. Pertinent comparisons are offered by experimental estimates of $ZPE + H_T - H_0$, namely, in kcal/mol, that of bicyclo[2.2.2]octane (130.83) [189] and those of cyclohexane (107.54), methylcyclohexane (125.73), $trans$-decalin (166.90), and adamantane (153.54), reported in Ref. 190, which were deduced from the spectroscopic data of Ref. 187 and the $H_T - H_0$ values given in Ref. 188.

The $(n_{cycle} + m - 1)RT/2$ term of Eq. (9.10) accounts for the fact that internal rotations are hindered in cyclic structures [44,180]: it was found that an amount of $\frac{1}{2}RT$ should be subtracted from $ZPE + H_T - H_0$ for each of the $(n_{cycle} + m - 1)$ carbon–carbon bonds making up the cycle(s). A test carried out for 1-$trans$-2-cis-3-trimethylcyclohexane does not suggest that the rotation of the methyl group on C-2 is hindered, according to the present results.

Alkenes

For the simple olefins, C_nH_{2n}, Eq. (9.11) (in kcal/mol)

$$ZPE + (H_T - H_0) = 33.35 + 18.213(n_C - 2) - 0.343n_{br} \qquad (9.11)$$

represents a valid approximation [44]. (Situations of extreme crowding, such as those arising with two $tert$-butyl groups attached to the same carbon, do not obey this equation.)

Let us now proceed with conjugated and nonconjugated polyenic hydrocarbons. Their $ZPE + (H_T - H_0)$ energies (Table 9.2) can be estimated [191] from those of their olefinic fragments [Eq. (9.11)] simply by subtracting 11.56 kcal/mol for each pair of hydrogen atoms eliminated in condensation of the fragments. For example, the result predicted for 1,3-pentadiene follows from the values of ethene and propene less 11.56 kcal/mol. Similarly, the result for 1,3,5-hexatriene corresponds to 3 times that of ethene less twice 11.56 kcal/mol or, alternatively, to the sum obtained from ethene and butadiene less 11.56 kcal/mol. Now, taking the experimental value of butadiene (55.19 kcal/mol) as reference, we can describe the dienes as follows [192], in kcal/mol:

$$ZPE + (H_T - H_0) = 55.19 + 18.213(n_C - 4) - 0.343n_{br} \qquad (9.12)$$

While this type of estimate usually carries an uncertainty not exceeding ~ 0.2 kcal/ mol, it remains that for the dienes (as for the monoolefins) no spectroscopic

TABLE 9.2. Zero-Point and Heat Content Energies of Olefins and Polyenes, at 298.15 K (kcal/mol)

Molecule	ZPE + $(H_T - H_0)$	
	Predicted	Experimental
Ethene	33.35	33.36
Propene	51.56	51.82
1-Butene	69.78	69.73
cis-2-Butene	69.78	69.91
trans-2-Butene	69.78	69.70
Isobutene	69.44	69.57
1,3-Butadiene	55.14	55.19
trans-1,3-Pentadiene	73.35	73.28
cis-1,3-Pentadiene	73.35	73.32
Isoprene	73.01	72.85
Dimethyl-1,3-butadiene	90.88	91.08
trans-1,3,5-Hexatriene	76.93	76.70
cis-1,3,5-Hexatriene	76.93	76.99
trans,trans-1,3,5,7-Octatetraene	98.72	99.97

Source: Ref. 44, which cites the sources of the frequencies used in the evaluation of the experimental values.

information is presently available that discriminates between ZPE + $(H_T - H_0)$ energies of *cis* and *trans* isomers in a reliable manner.

Aromatic Hydrocarbons

Spectroscopic vibrational data of benzenoid hydrocarbons are scarce. Fortunately, it is now justifiable to take advantage of the regularities of the ZPE + $(H_T - H_0)$ energies observed during the buildup of alkyl chains, namely, the gain of 18.213 kcal/mol for each added CH_2 group. Hence it appears safe to use the following formula for alkyl substitution, based on the experimental ZPE + $(H_T - H_0)$ value (66.22 kcal/mol) deduced for benzene in the harmonic oscillator approximation [27,193]

$$ZPE + (H_T - H_0) = 66.22 + 18.21n - 0.343n_{br} \qquad (9.13)$$

where n is the number of the alkyl carbon atoms. One can proceed in similar fashion with molecules like 1,2,3,4-tetrahydronaphthalene **1**, or 9,10-dihydroanthracene **2**:

1   **2**

However, in these situations one must account for the lowering of ZPE $+ (H_T - H_0)$ by 11.5 kcal/mol accompanying the loss of two hydrogen atoms and by $RT/2 = 0.296$ kcal/mol for each hindered internal rotation in cyclic structures. 1,2,3,4-Tetrahydronaphthalene, for example, gives $66.22 + 4 \times 18.21 - 11.5 - 5 \times 0.296 = 126.1$ kcal/mol. The addition of two fragments involves a correction of $4 \times RT = 2.37$ kcal/mol because this term is included in the $(H_T - H_0)$ part of each molecule used as fragment and should not be counted twice in the final sum. For example, the ZPE $+ (H_T - H_0)$ value of 9,10-dihydroanthracene is estimated from two benzene molecules plus two CH_2 groups $(2 \times 66.22 + 2 \times 18.21)$ less 2×11.5, less $4RT$, and less $4 \times RT/2$. These estimates are considered to carry an uncertainty not exceeding 0.2 kcal/mol [129].

The additivity rules described in Ref. 129 can be applied with confidence to construct the ZPE $+ (H_T - H_0)$ energies of polycyclic aromatic hydrocarbons. For example, using the result ZPE $+ (H_T - H_0) = 94.90$ kcal/mol for naphthalene, deduced from its vibrational spectrum [194], we add to it the difference 28.68 kcal/mol between naphthalene and benzene, thus obtaining 123.6 kcal/mol for anthracene. The same procedure is used for the higher homologs. Finally, using the fundamental frequencies of pyrene [195], it is found that ZPE $+ (H_T - H_0) = 133.05$ kcal/mol. For styrene, one obtains 85.80 kcal/mol from its vibrational spectrum [196]; estimated, 85.71 kcal/mol.

As for graphite, its zero-point energy, ZPE $= \frac{2}{3}R\theta_{\parallel} + \frac{1}{4}R\theta_{\perp}$, is most conveniently deduced from Debye's theory [197,198] by separating the lattice vibrations into two approximately independent parts, with Debye temperatures $\theta_{\parallel}$ (in plane) and $\theta_{\perp}$ (perpendicular). A balanced evaluation gives ZPE $\simeq 3.68$ kcal/mol [199].

Amines and Hydrazines

The ZPE $+ (H_T - H_0)$ energies of amines and hydrazines also obey simple additivity rules [191] (Table 9.3). For the amines, one can predict their ZPE $+ (H_T - H_0)$ energies from that of NH_3 (22.94 kcal/mol) and the appropriate alkane, Eq. (9.9), by subtracting 11.55 kcal/mol for each pair of H atoms lost during the condensation, giving the amine. Dimethylamine, for example, is estimated by taking twice the value of CH_4 plus that of NH_3, less twice 11.55 kcal/mol. We proceed in similar fashion with the alkyl-substituted hydrazines.

Carbonyl Compounds and Ethers

Complete vibrational analyses of carbonyl compounds are scarce. Data obtained from experimental and calculated fundamental frequencies of acetaldehyde, acetone, and diethylketone are indicated in Table 9.4, in the harmonic oscillator approximation, for $T = 298.15$ K.

An extensive analysis [27] involving theoretical ΔE_a^* energies and experimental enthalpies of formation [cf. Eq. (9.6)] indicates that an increment of ~ 18.3 kcal/mol can be associated with each added CH_2 group, with respect to the closest parent compound whose spectroscopic ZPE $+ (H_T - H_0)$ result is known—which

TABLE 9.3. Zero-Point and Heat Content Energies^a of Amines and Hydrazines, at 298.15 K (kcal/mol)

Molecule	ZPE + $(H_T - H_0)$	
	Predicted	Experimental
Ammonia	(22.94)	22.94
Methylamine	41.08	41.48
Ethylamine	59.30	59.41
Propylamine	77.51	77.62
Isopropylamine	77.17	77.20
tert-Butylamine	95.38	95.21
Dimethylamine	59.22	59.44
Diethylamine^b	95.65	94.94
Trimethylamine	77.37	77.76
Hydrazine	(33.94)	33.94
Methylhydrazine	52.08	52.00
1,1-Dimethylhydrazine	69.88	69.98
1,2-Dimethylhydrazine	70.22	70.20

^{a}From Ref. 44, which cites the sources of the frequencies used in the evaluation of the experimental values.
^{b}No provision was made for the (at least partial) hindrance of internal rotations in the ethyl groups as a result of steric crowding, suggesting that our estimate is therefore too high by ~0.6 kcal/mol.

seems reasonable. For example, the value for propanal (36.65 + 18.3) is estimated from that of ethanal (36.65 kcal/mol), and the value for butanone (72.89) is estimated from that of propanone (54.59 kcal/mol). Energy calculations, namely, those of standard enthalpies of formation, carried out with the help of these estimates are in excellent agreement with experimental results.

TABLE 9.4. ZPE + $(H_T - H_0)$ Energies of Carbonyl Compounds and Ethers, at 298.15 K (kcal/mol)

Molecule	ZPE + $(H_T - H_0)$	
	Estimated	Experimental
CH_3CHO	—	36.65
$(CH_3)_2CO$	—	54.59
$(C_2H_5)_2CO$	—	91.52
$(CH_3)_2O$	52.55	52.65
$CH_3OC_2H_5$	70.19	69.86
CH_3O i-C_3H_7	87.83	87.66
$(C_2H_5)_2O$	87.83	87.76
C_2H_5O i-C_3H_7	105.47	105.40
$(i$-$C_3H_7)_2O$	123.11	123.32

The dialkylethers are adequately described by the approximation [27] (in kcal/mol)

$$\text{ZPE} + (H_T - H_0) = 52.55 + 17.64(n_C - 2) \tag{9.14}$$

with an average deviation between predicted and experimental results of $\sim$0.16 kcal/mol (Table 9.4).

Free Radicals

Regarding the $\text{ZPE} + H_T - H_0$ energies, little is presently known for the free radicals, except for a few alkyl radicals. Relevant $\text{ZPE} + H_T - H_0$ energies, deduced from both experimental and calculated fundamental frequencies [200–202], are 20.74 ($CH_3\cdot$), 39.15 ($C_2H_5\cdot$), and 74.97 kcal/mol (*tert*-$C_4H_9\cdot$). These results suggest that the $\text{ZPE} + H_T - H_0$ energies of alkyl radicals $R\cdot$ are systematically lower by $\sim$8.85 kcal/mol than those of the parent hydrocarbons RH. Additional results were obtained by standard methods in the harmonic oscillator approximation. Selected values for $\text{ZPE} + H_T - H_0$ are, in kcal/mol, 12.52 (CH_2 [203]), 13.83 ($NH_2\cdot$ [204]), 23.20 ($CH_2{=}CH\cdot$ [205]), 31.78 ($CH_3NH\cdot$ [139]), and 32.63 ($NH_2CH_2\cdot$ [139]).

9.3 CONCLUDING REMARKS

Simple additivity rules relating $\text{ZPE} + (H_T - H_0)$ energies to structural features have proved their usefulness in the past.

Here they are examined primarily because they greatly facilitate the comparison between thermochemical results and calculations made for molecules in their hypothetical vibrationless state at 0 K. While, of course, preference is given to verifications involving only genuine experimental data, well-established structure-dependent regularities of zero-point plus heat content energies considerably augments the number of molecules that can be tested.

At times, the quality of the correlations obtained from bona fide spectral analyses surpasses what one would normally expect. In cycloalkanes, for example, any shrinking of a cycle accompanying the removal of one CH_2 group translates into a regular decrease of $\text{ZPE} + (H_T - H_0)$ by 18.545 kcal/mol; it is truly remarkable that this regularity includes the shrinking of cyclopropane to give the "two-membered cycle" ethylene.[1]

In general, things are simpler than that, much to our advantage. Within the limits set by the precision of the present estimates, structural features like the chair, boat, or twist–boat conformations of cyclohexane rings, as well as the butane-*gauche* effects or the *cis–trans* isomerism of ethylenic compounds leave no recognizable distinctive trace in zero-point plus heat content energies. Indeed, whatever residual, presently

---

[1]The predicted $\text{ZPE} + (H_T - H_0)$ values are, in kcal/mol (those derived from experimental frequencies are in parentheses), cycloheptane, 126.09 (124.15); cyclohexane, (107.54); cyclopentane, 89.00 (89.0); cyclobutane, 70.45 (70.4); cyclopropane, 57.91 (57.9); and ethylene, 33.36 (33.36) [44].

unavoidable uncertainties remain attached to these quantities, they could hardly be blamed for anything as they are unlikely to impair the validity of comparisons between theory and experimental values, which, by the way, are also affected by well-known margins of errors.

Finally, it is unfortunate that the poor harvest of pertinent data for free radicals limits our means in an important area of chemistry, that concerned with the making and breaking of chemical bonds, because the way the $ZPE + (H_T - H_0)$ energy changes in going from a molecule to its fragments (or in the reverse process) is a relevant part of the energy balance accompanying chemical reactions.

CHAPTER 10

THE CHEMICAL BOND: THEORY (I)

10.1 SYNOPSIS

Let us first sketch out the conceptual content of our bond energy method [44,108]. For a fair discussion of chemical binding, molecules are best considered in their hypothetical vibrationless state at 0 K when entering the reaction

$$\text{Molecule} \longrightarrow \text{ground-state atoms}$$

The energy of this atomization (in conventional notation)

$$\Delta E_a^* = \sum_k \langle \Psi_k | \hat{H}_k^{at} | \Psi_k \rangle - \langle \Psi^{mol} | \hat{H}^{mol} | \Psi^{mol} \rangle \tag{10.1}$$

$$= \sum_{k<l} \varepsilon_{kl} - E_{nb} \tag{10.2}$$

is taken as a sum of bond energy contributions ε_{kl} between atoms k and l, and a change of nonbonded energy terms $\Delta E_{nb} = -E_{nb}$, which accompanies atomization. The latter part is very small, indeed, say, of the order of $0.01-0.05\%$ of ΔE_a^* and is adequately evaluated in the point charge approximation [206]

$$E_{nb} = \frac{1}{2} \sum_{r,s}^{nb} \frac{q_r q_s}{R_{rs}} \tag{10.3}$$

Atomic Charges, Bond Properties, and Molecular Energies, by Sándor Fliszár
Copyright © 2009 John Wiley & Sons, Inc.

which suffices in our intended applications. Leaving this subject momentarily, we focus attention on the bonded part, $\sum_{k<l} \varepsilon_{kl}$.

The mental decomposition of ΔE_a^* into bond energy terms and the description of the latter featuring the atomic charges of the bond-forming atoms rests on three main ideas:

- Using Eq. (10.1) to obtain, with the help of the Hellmann–Feynman theorem [74], the derivative of ΔE_a^* at constant electron density ρ, taking the nuclear charge Z_k as variable, namely

$$\left(\frac{\partial \Delta E_a^*}{\partial Z_k}\right)_\rho = \frac{V_{ne,k}}{Z_k} - \frac{V_k}{Z_k} \tag{10.4}$$

where $V_{ne,k}/Z_k$ is the potential at the nuclear position in the isolated atom and V_k/Z_k is the potential at the nucleus Z_k in the molecule. The problem has thus evolved into a simple electrostatic problem of potential energies where the interelectronic repulsions, as well as the kinetic energy, which are not functions of Z_k, have disappeared.

- Using the Thomas–Fermi approximation

$$E^{molecule} = \frac{1}{\gamma}\sum_k V_k$$

which is well documented [44,79,207]. This formula enables us to account for both the interelectronic repulsion terms and the electronic kinetic energy although only nuclear–electronic and internuclear potential energies are considered explicitly in V_k.

- Applying Gauss' theorem, leading to the Politzer–Parr core–valence separation in atoms [61]

$$E^{valence} = -\frac{1}{\gamma^v} Z^{eff} \int_{r_b}^\infty \frac{\rho(r)}{r}\, d\mathbf{r} \tag{10.5}$$

where $Z^{eff} = Z - N^c$ is the effective nuclear charge (e.g., 4 for carbon, 5 for nitrogen [44]), that is, the nuclear charge minus the number of core electrons. The integral is carried out beyond the boundary r_b that delimits the region containing the N^c core electrons. This simplification conveys the physical picture of a valence electron cloud in the field of a nucleus partially screened by its core electrons and is a form of the Thomas–Fermi approximation. The integral can be written $N\langle r^{-1}\rangle$, where N is the number of valence electrons and $\langle r^{-1}\rangle$ is the expectation value of r^{-1}. Hence,

$$E^{valence} = -\frac{1}{\gamma^v} Z^{eff} N\langle r^{-1}\rangle \tag{10.6}$$

Note that γ and γ^v usually approach the Thomas–Fermi limit $\left(\frac{7}{3}\right)$ (Table 4.4) except for hydrogen, where $\gamma = 2$ because of the virial theorem, $E = T + V_{ne}$, with $E = -T$.

Recall that the core–valence separation in molecules is described in real space [83], as any atom-by atom or bond-by-bond partitioning of a molecule is inherently a real-space problem. Equation (10.6) does indeed refer to a *partitioning in real space* (as opposed to the usual *Hartree–Fock orbital space*), both for ground-state isolated atoms or ions and for atoms embedded in a molecule, with $N^c = 2\,e$ for first-row elements.

This concludes the enumeration of the concepts involved in our bond energy theory.

Now we turn our attention to the intrinsic bond energies, ε_{kl}. Two approaches will be considered: (1) a derivation making use of Eq. (10.2) and (2) a derivation rooted in Eq. (4.47). But first, we deal with the nonbonded part, Eq. (10.3), and examine future possible simplifications regarding this energy contribution.

10.2 NONBONDED INTERACTIONS

The idea behind this survey of nonbonded interactions is to get rid of them elegantly as explicit terms requiring separate calculations by means of Eq. (10.3). We shall examine to what extent nonbonded Coulomb-type interactions are at least approximately additive. The formulation of additivity is presented here for $C_nH_{2n+2-2m}$ hydrocarbons [208], where m is the number of six-membered cycles.

Let X be a molecular property (e.g., E_{nb}), and let $X^{C_2H_6}$, X^{CH_4} be the corresponding values for ethane and methane, respectively. If X is exactly additive, then

$$X = (1 - m)X^{C_2H_6} + (n - 2 + 2m)\left(X^{C_2H_6} - X^{CH_4}\right) \qquad (10.7)$$

where $X^{C_2H_6} - X^{CH_4}$ is the change in X on going from methane to ethane, namely, the contribution of one CH_2 group. The meaning of Eq. (10.7) is obvious for acyclic molecules ($m = 0$). For example, the X value for propane is that of ethane plus the increment corresponding to one added CH_2 group. For cyclohexane ($m = 1$), which consists of $n - 2 + 2m = 6$ CH_2 groups, the $(1 - m)X^{C_2H_6}$ term of (10.7) cancels.

Decalin is constructed from two cyclohexane units. In this case the second RHS term of Eq. (10.7), $(n - 2 + 2m)(X^{C_2H_6} - X^{CH_4})$, accounts for 12 CH_2 groups, but one additional $X^{C_2H_6}$ contribution (i.e., that of two CH_2 and two H atoms) is subtracted with respect to cyclohexane, that is, a total of two $X^{C_2H_6}$ contributions with respect to acyclic alkanes. Similar arguments applied to other polycyclic saturated hydrocarbons verify the validity of Eq. (10.7) as a formulation of exact additivity.

The results indicated in Table 10.1 were deduced from Eq. (10.3) using charges corresponding to the scale defined by $q_C^{ethane} = 35.1$ me. Attractive (i.e., stabilizing) interactions are negative. Of course, it is not surprising that branching favors repulsive contributions.

TABLE 10.1. Nonbonded Coulomb Interactions

Molecule	E_{nb} (kcal/mol)
1 Methane	0.09
2 Ethane	−0.07
3 Propane	−0.20
4 Butane	−0.32
5 Isobutane	−0.28
6 Pentane	−0.44
7 Isopentane	−0.39
8 Neopentane	−0.32
9 2,2-Dimethylbutane	−0.43
10 2,3-Dimethylbutane[a]	−0.39
11 2,2,3-Trimethylbutane	−0.46
12 2,2,3,3-Tetramethylbutane	−0.44
13 Cyclohexane	−0.73
14 Bicyclo[2.2.2]octane	−1.03
15 Bicyclo[3.3.1]nonane	−1.13
16 *trans*-Decalin	−1.26
17 *cis*-Decalin	−1.23
18 Adamantane	−1.31
19 Iceane	−1.61

[a]Calculated for the statistical average of one *anti* (−0.40) and two *gauche* (−0.39) forms, as discussed in Ref. 209.

It turns out that the nonbonded terms of saturated hydrocarbons are approximately additive. From Eq. (10.7), the appropriate formulation is thus

$$E_{nb} \simeq (1 - m)E_{nb}^{C_2H_6} + (n - 2 + 2m)\left(E_{nb}^{C_2H_6} - E_{nb}^{CH_4}\right) \qquad (10.8)$$

if we agree on accepting errors of $\sim$0.1 kcal/mol due to the neglect of differences between isomers. This equation, applied to the energies given in Table 10.1, yields the result presented in Fig. 10.1. This result is self-explanatory.

It appears, indeed, that nonbonded Coulomb interactions behave in general in a "quasiadditive" manner in terms of Eq. (10.7) as long as we neglect the (minor) differences between isomers. Branching causes a systematic trend toward higher energies (repulsive destabilization), but situations of extreme crowding are required, such as those encountered in 2,2,3,3-tetramethylbutane (**12**) and, to a lesser extent, in 2,2,3-trimethylbutane (**11**), in order to produce sizeable departures from "quasiadditivity," that is, from Eq. (10.8).

With these reservations in mind, a simple way of taking advantage of the "quasiadditivity" of nonbonded interactions in saturated hydrocarbons is implemented in Section 10.6. The unimportant loss in precision is largely justified by the considerable simplification thus achieved in calculations of atomization energies.

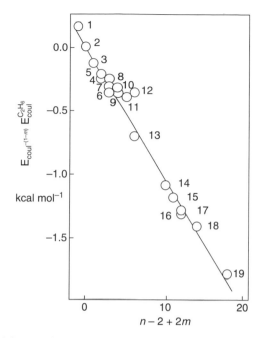

Figure 10.1. Additivity test for Coulomb nonbonded interaction energies, Eq. (10.8). The radii of the degles represent an uncertainty of $\sim 0.03\,\text{kcal/mol}$. The points correspond to the numbering used in Table 10.1. (From Ref. 207.)

10.3 REFERENCE BONDS

Equation (10.2) is our starting point. Applying the Hellmann–Feynman theorem, we get

$$\left(\frac{\partial \Delta E_a^*}{\partial Z_k}\right)_\rho = \sum_l \left(\frac{\partial \varepsilon_{kl}}{\partial Z_k}\right)_\rho - \left(\frac{\partial E_{nb}}{\partial Z_k}\right)_\rho \tag{10.9}$$

Hence, with the help of Eq. (10.4), we obtain

$$V_k = V_{ne,k} - Z_k \sum_l \left(\frac{\partial \varepsilon_{kl}}{\partial Z_k}\right)_\rho + Z_k \left(\frac{\partial E_{nb}}{\partial Z_k}\right)_\rho \tag{10.10}$$

Now we use Eq. (3.21) with the appropriate labels (k, l, etc.), for instance, $E_k^{at} = (1/\gamma_k^{at})V_{ne,k}$, as well as (4.10), $E_k^{mol} = (1/\gamma_k^{mol})V_k$, and calculate ΔE_k with the help of Eq. (4.13). This gives

$$\Delta E_k = \frac{Z_k}{\gamma_k^{mol}} \sum_l^{bd} \left(\frac{\partial \varepsilon_{kl}}{\partial Z_k}\right)_\rho + \frac{\gamma_k^{mol} - \gamma_k^{at}}{\gamma_k^{mol}} E_k^{at} - \frac{Z_k}{\gamma_k^{mol}} \left(\frac{\partial E_{nb}}{\partial Z_k}\right)_\rho \tag{10.11}$$

ΔE_k measures the binding of atom k in the molecule. It is obvious that the last term of the right-hand side of Eq. (10.11) has nothing to do with the chemical bonds formed by that atom. Only the first and second terms are involved in the decomposition of ΔE_k among the bonds formed by atom k. The "extraction" from the host molecule of atom k forming v_k bonds requires an energy

$$\frac{1}{v_k} \cdot \frac{\gamma_k^{mol} - \gamma_k^{at}}{\gamma_k^{mol}} E_k^{at}$$

for each individual atom, meaning that in the suppression of a kl bond this type of contribution must be counted once for atom k and once for atom l. In addition, the suppression of the kl bond requires an energy $(1/\gamma_k^{mol})Z_k(\partial \varepsilon_{kl}/\partial Z_k)_\rho$ and a similar energy for the partner l engaged in that bond. Consequently, the portion of the total atomization energy associated with the kl bond is

$$\varepsilon_{kl} = \frac{Z_k}{\gamma_k^{mol}} \left(\frac{\partial \varepsilon_{kl}}{\partial Z_k}\right)_\rho + \frac{Z_l}{\gamma_l^{mol}} \left(\frac{\partial \varepsilon_{kl}}{\partial Z_l}\right)_\rho$$

$$+ \frac{1}{v_k} \cdot \frac{\gamma_k^{mol} - \gamma_k^{at}}{\gamma_k^{mol}} E_k^{at} + \frac{1}{v_l} \cdot \frac{\gamma_l^{mol} - \gamma_l^{at}}{\gamma_l^{mol}} E_l^{at} \qquad (10.12)$$

which is the sought-after expression for the intrinsic energy of a chemical bond [108]. Bond energies given by Eq. (10.12) satisfy exactly Eq. (10.2).

Direct applications of Eq. (10.12) are generally difficult to handle—this is why the more efficient charge-dependent energy formulas were developed in the first place. Most thorough tests were made for selected carbon–carbon bonds [13,14,44,108] (Table 1.1).

But there are ways to circumvent the difficulty of solving Eq. (10.12), at least in an approximate manner. Consider the Hellmann–Feynman derivative (10.4) for a hydrogen atom bonded to atom l

$$V_H = V_{ne,k} - Z_k \left(\frac{\partial \varepsilon_{Hl}}{\partial Z_k}\right)_\rho \qquad (10.13)$$

with $V_{ne,k} = -1$ au for the isolated H atom and where V_H is (for $k =$ hydrogen) the classical electrostatic potential energy

$$V_k = -Z_k \int \frac{\rho(\mathbf{r})}{|\mathbf{r} - \mathbf{R}_k|} d\mathbf{r} + Z_k \sum_{l \neq k} \frac{Z_l}{R_{kl}} \qquad (10.14)$$

A *small* change of ε_{kl}, written $\Delta \varepsilon_{kl}$, can be evaluated in an approximate manner by replacing the first term of Eq. (10.12) with a modified one, leaving the rest of Eq. (10.12) unchanged, so that (with $\gamma_H^{mol} = 2$)

$$\Delta \varepsilon_{kl} = \frac{Z_k}{\gamma_k^{mol}} \left[\left(\frac{\partial \varepsilon_{kl}}{\partial Z_k}\right)_\rho^{modified} - \left(\frac{\partial \varepsilon_{kl}}{\partial Z_k}\right)_\rho^{original} \right]$$

Thus, with the help of Eq. (10.13), for hydrogen linked to atom l, we have

$$\Delta\varepsilon_{Hl} = -\frac{1}{2}\left(V_H^{\text{modified}} - V_H^{\text{original}}\right) \tag{10.15}$$

This is one of our working formulas. It is an approximation, of course, but we are presently unable to do better: evidently, by implementing this approximation, we possibly transfer additional contingent variations of the ε_{Hl} bond energy to other bonds formed by atom l in the molecule. With this reservation in mind, we shall illustrate the use of Eq. (10.15) in Chapter 15, thus revealing instructive bond properties.

Summa summarum, Eq. (10.12) is certainly not the most efficient one for routine calculations of intrinsic bond energies. The reason is obvious—it lies with the partial derivatives $(\partial\varepsilon_{kl}/\partial Z_k)_\rho$, which must be carried out at constant electron density ρ, meaning that this difficult calculation has to be made for each new ε_{kl}, which is unpractical.

Fortunately, we have something better in store. First we calculate a few reference bond energies ε_{kl}°, using Eq. (10.12), and subsequently modify these bond energies as the electron densities ρ are varied.

Example 10.1: The CC and CH Bonds of Ethane. Approximate solutions can be found in some favorable cases with the help of Eq. (10.10), as indicated in the following example worked out for ethane.

First we apply Eq. (10.14). SCF results for methane and ethane are[1] $V_C^{CH_4} = -88.52015$, $V_H^{CH_4} = -1.12634$, $V_C^{C_2H_6} = -88.45999$, and $V_H^{C_2H_6} = -1.13355$ au, with total energies $E^{\text{mol}} = -40.2090$ and -79.2513 au, respectively. Next we apply $E^{\text{mol}} = \sum_k (1/\gamma_k)V_k$, from Eqs. (4.10)–(4.12), assuming that the SCF γ_k^{mol} parameters are valid for real molecule and rescale the SCF potential energies to reproduce the corresponding energies of atomization; so we get the *rescaled values* $V_C^{CH_4} = -89.2305$, $V_H^{CH_4} = -1.1354$, $V_C^{C_2H_6} = -89.1270$, and $V_H^{C_2H_6} = -1.1421$ au. At last we can estimate the CH and CC bond energies of ethane. This can be done in two ways.

Method 1. Remembering that $V_{ne} = -1$ au for the H atom, Eq. (10.10) tells us that, to a good approximation, $(\partial\varepsilon_{CH}/\partial Z_H)$ is 0.1354 au for methane and 0.1421 au for ethane. Now consider Eq. (10.12) and suppose that $(\partial\varepsilon_{CH}/\partial Z_C)$ is at least approximately the same in both molecules.[2] In this approximation, Eq. (10.12) tells us that the ε_{CH} energy of ethane is larger than that of methane by $\sim\frac{1}{2}(0.1421 - 0.1354)$ au = 2.10 kcal/mol. Since the latter is ~104.81 kcal/mol, we find for ethane that $\varepsilon_{CH}^\circ \simeq 106.91$ and thus $\varepsilon_{CC}^\circ \simeq 69.1$ kcal/mol, with reference to the experimental energies of methane and ethane, $\Delta E_a^* = 419.24$ and 710.54 kcal/mol, respectively.

---

[1]Pople's 6-311G** basis [210] and experimental geometries [211] of methane ($R_{CH} = 1.085$ Å) and ethane ($R_{CC} = 1.531$ Å, $R_{CH} = 1.096$ Å, and $\angle HCH = 107.8°$) were used.
[2]This unavoidable hypothesis, under the present degumstances, seems justifiable because the charge difference between these carbons is very small.

TABLE 10.2. Selected SCF γ_k^{at} and γ_k^{mol} Parameters

Atom	γ_k^{at}	γ_k^{mol}
H	2	2
B	2.31965	2.31428
C	2.33863	2.33263
N	2.35905	2.34742
O	2.38039	2.37361
F	2.40096	2.39866

Method 2. Here we use $V_C^{at} = -88.5307$ au for carbon, deduced from its experimental energy, -37.8558 au and $\gamma_C^{at} = 2.33863$ (see Table 10.2). Equation (10.10) gives, in atomic units, $-89.2305 = -88.5307 - Z_C[4(\partial\varepsilon_{CH}/\partial Z_C)]$ for methane and $-89.1270 = -88.5307 - Z_C[3(\partial\varepsilon_{CH}/\partial Z_C) + (\partial\varepsilon_{CC}/\partial Z_C)]$ for ethane. Now, assuming as before the same value of $(\partial\varepsilon_{CH}/\partial Z_C)$ for the two molecules, we get $(\partial\varepsilon_{CH}/\partial Z_C) = 0.02916$ and $(\partial\varepsilon_{CC}/\partial Z_C) = 0.0119$ au. Equation (10.12) thus gives $\varepsilon_{CC}^\circ = 69.0$ kcal/mol.

Note that the first calculation uses only the difference between the potential energies $V_H^{C_2H_6}$ and $V_H^{CH_4}$ to get the CH bond energy of ethane from that of methane, which is known. The second calculation involves $V_C^{C_2H_6}$, $V_C^{CH_4}$, and V_C^{atom} *and*, most importantly, depends strongly on the γ parameters required for solving (10.12). Under these difficult circumstances, it is gratifying that both methods give approximately the same result for ε_{CC}°. The calculations of ε_{CH}° and ε_{CC}° using Eq. (10.12) suffer from imprecisions and can do no better than suggesting an approximate value, ~69 kcal/mol for the CC bond energy in ethane. The final selection is described in Section 10.6.

10.4   BOND ENERGY: WORKING FORMULAS

Now we go back to Eq. (4.47), namely

$$\Delta E_a^* = \Delta E_a^{*\circ} - \frac{1}{\gamma^v}\Delta\left(V_{ne}^{eff} + 2V_{nn}^{eff}\right)$$

and write ε_{kl}° for a reference bond energy corresponding to the electron density ρ° for that reference, while ε_{kl} corresponds to a modified density. From now on, the electron densities ρ and the electron populations N refer to valence electrons. The $(1/\gamma^v)\Delta(V_{ne}^{eff} + 2V_{nn}^{eff})$ term concerns only the interactions between bonded atoms.[3] We shall momentarily assume (and this is quite an assumption) that we know how to assign the electron populations N_k, N_l, ... to the individual "atoms in a molecule," k, l, and so on.

---

[3]Of course, any modification of $V_{ne}^{eff} + 2V_{nn}^{eff}$ affects the nonbonded part as well, but this effect is automatically included in the nonbonded contribution calculated for the density ρ: no problem arises, owing to the almost negligible weight of E_{nb}.

The contribution to ΔV_{ne}^{eff} involving Z_k^{eff} consists of the interactions between Z_k^{eff} and the electrons with density ρ_k that are assigned to atom k in the molecule, namely

$$- Z_k^{eff} \int^{\tau_k} \frac{\rho_k(\mathbf{r})}{|\mathbf{r} - \mathbf{R}_k|} \, d\mathbf{r} \tag{10.16}$$

where the integration is carried out over the volume τ_k containing the N_k electrons allocated to k:

$$N_k = \int^{\tau_k} \rho_k(\mathbf{r}) \, d\mathbf{r}$$

On the other hand, Z_k interacts with the N_l electrons of each atom l bonded to k

$$- Z_k^{eff} \int^{\tau_l} \frac{\rho_l(\mathbf{r})}{|\mathbf{r} - \mathbf{R}_k|} \, d\mathbf{r} \tag{10.17}$$

with

$$N_l = \int^{\tau_l} \rho_l(\mathbf{r}) \, d\mathbf{r}$$

The integrals in (10.16) and (10.17) are conveniently written

$$\int^{\tau_k} \frac{\rho_k(\mathbf{r})}{|\mathbf{r} - \mathbf{R}_k|} \, d\mathbf{r} = N_k \langle r_k^{-1} \rangle \tag{10.18}$$

$$\int^{\tau_l} \frac{\rho_l(\mathbf{r})}{|\mathbf{r} - \mathbf{R}_k|} \, d\mathbf{r} = N_l \langle r_{kl}^{-1} \rangle \tag{10.19}$$

where $\langle r_k^{-1} \rangle$ and $\langle r_{kl}^{-1} \rangle$ are the average inverse distances from Z_k^{eff} to N_k and N_l, respectively. The contribution involving Z_k^{eff} and its "own" electrons N_k, on one hand, and the electrons N_l of every atom l forming a bond with k, on the other hand, is as follows, using (10.16)–(10.19):

$$V_{ne,k}^{eff} = -Z_k^{eff} \left[N_k \langle r_k^{-1} \rangle + \sum_l N_l \langle r_{kl}^{-1} \rangle \right] \tag{10.20}$$

Here we specify that this expression holds for the true densities of the system under scrutiny. Similar expressions are written for the reference model with atomic populations N_k° and N_l° and average inverse distances $\langle r_k^{-1} \rangle^\circ$ and $\langle r_{kl}^{-1} \rangle^\circ$. This gives an equation like (10.20), but for model densities $\rho^\circ(\mathbf{r})$. Hence we obtain the difference $\Delta V_{ne,k}^{eff} = V_{ne,k}^{eff} - V_{ne,k}^{eff}(\rho^\circ)$ for nucleus Z_k^{eff}. Summation over all nuclei and inclusion of the $2\Delta V_{nn}^{eff}$ term gives

$$\Delta(V_{ne}^{eff} + 2V_{nn}^{eff}) = - \sum_k Z_k^{eff} \left[N_k \langle r_k^{-1} \rangle - N_k^\circ \langle r_k^{-1} \rangle^\circ + \sum_l (N_l \langle r_{kl}^{-1} \rangle - N_l^\circ \langle r_{kl}^{-1} \rangle^\circ) \right]$$

$$+ \sum_k \sum_l Z_k^{eff} Z_l^{eff} \left[R_{kl}^{-1} - (R_{kl}^{-1})^\circ \right] \tag{10.21}$$

What this equation says is simply that the addition of a small amount of electronic charge to an atom k modifies its "own" atomic nuclear–electronic interaction energy, namely, that involving Z_k, and, moreover, also changes the nuclear–electronic interactions with the nuclei of all atoms l bonded to k. The effects involving the remaining atoms of the molecule are, of course, included in the small nonbonded interaction term, E_{nb}. Note that this presentation of $\Delta(V_{ne}^{eff} + 2V_{nn}^{eff})$ does not imply acceptance of the point-charge model, although E_{nb} itself will ultimately be evaluated in the point-charge approximation.

Let us transform Eq. (10.21) into something more practical. First, we replace $\langle r_{kl}^{-1} \rangle$ by $\langle r_{kl}^{-1} \rangle^{\circ}$ where it occurs in (10.21) and restore the correct result in the following manner:

$$
\Delta(V_{ne}^{eff} + 2V_{nn}^{eff}) = -\sum_k Z_k^{eff}\left[N_k \langle r_k^{-1} \rangle - N_k^{\circ}\langle r_k^{-1}\rangle^{\circ} + \sum_l \left(N_l \langle r_{kl}^{-1}\rangle^{\circ} - N_l^{\circ}\langle r_{kl}^{-1}\rangle^{\circ}\right)\right]
$$

$$
+ \sum_k \sum_l Z_k^{eff} Z_l^{eff}\left[R_{kl}^{-1} - (R_{kl}^{-1})^{\circ}\right]
$$

$$
- \sum_k \sum_l Z_k^{eff}\left(N_l \langle r_{kl}^{-1}\rangle - N_l \langle r_{kl}^{-1}\rangle^{\circ}\right) \tag{10.22}
$$

At this point we rewrite Eq. (4.47) as follows:

$$
\sum_{k<l} \varepsilon_{kl} = \sum_{k<l} \varepsilon_{kl}^{\circ} - \frac{1}{\gamma^v}\Delta\left(V_{ne}^{eff} + 2V_{nn}^{eff}\right) \tag{10.23}
$$

This means that the reference bond energies, ε_{kl}° corresponding to a reference electron density $\rho^{\circ}(\mathbf{r})$ have been modified by the change $\Delta(V_{ne}^{eff} + 2V_{nn}^{eff})$ to give the energies ε_{kl} corresponding to $\rho(\mathbf{r})$. We also define

$$
N_l = N_l^{\circ} + \Delta N_l \tag{10.24}
$$

for use in Eq. (10.22) and thus obtain from Eq. (10.23) that

$$
\sum_{k<l} \varepsilon_{kl} = \sum_{k<l} \varepsilon_{kl}^{\circ} + F + \frac{1}{\gamma^v}\sum_k Z_k^{eff}\left[\left(N_k \langle r_k^{-1}\rangle - N_k^{\circ}\langle r_k^{-1}\rangle^{\circ}\right) + \sum_l \Delta N_l \langle r_{kl}^{-1}\rangle^{\circ}\right]
$$

$$
\tag{10.25}
$$

The function F consists of the last two terms of (10.22) divided by γ^v, with a change in sign. It is convenient to rewrite F using the definition of net atomic charge, $Z_l^{eff} - N_l = q_l$. After some algebra one obtains

$$
F = -\frac{1}{\gamma^v}\sum_k \sum_l Z_k^{eff} Z_l^{eff}\left[R_{kl}^{-1} - (R_{kl}^{-1})^{\circ} - \left(\langle r_{kl}^{-1}\rangle - \langle r_{kl}^{-1}\rangle^{\circ}\right)\right]
$$

$$
- \frac{1}{\gamma^v}\sum_k \sum_l Z_k^{eff} q_l\left(\langle r_{kl}^{-1}\rangle - \langle r_{kl}^{-1}\rangle^{\circ}\right) \tag{10.26}
$$

F represents the contribution due to variations of internuclear distances, say, R_{kl} instead of the reference value R_{kl}°, and to changes of electronic centers of charge (i.e., of the distance from Z_k^{eff} to N_l) from r_{kl}° to r_{kl}. $F = 0$ proves accurate in most applications to σ systems.

The first part, in square brackets, is obviously 0 for spherically symmetric electron clouds and, more generally, if the centers of electronic charge move along with the nuclei during small changes in internuclear distances. The last part is small because atomic charges are small in the first place (e.g., ≤ 0.0351 e for carbon) and because small changes of electron populations are unlikely to modify their center of charge to any significant extent. The function F no longer equals zero when $C(sp^2)$—$C(sp^3)$ and $C(sp^2)$—H bonds are found in a molecule: F will be used to obtain the energies of these bonds from those of the parent saturated molecule. It gives means to deal sensibly with situations departing from the simple point-charge approximation.

Let us now consider the difference $N_k\langle r_k^{-1}\rangle - N_k^{\circ}\langle r_k^{-1}\rangle^{\circ}$ appearing in Eq. (10.25) and expand $N_k\langle r_k^{-1}\rangle$ as follows in a Taylor series:

$$N_k\langle r_k^{-1}\rangle = N_k^{\circ}\langle r_k^{-1}\rangle^{\circ} + \left(\frac{\partial N_k\langle r_k^{-1}\rangle}{\partial N_k}\right)^{\circ}\Delta N_k$$

$$+ \frac{1}{2!}\left(\frac{\partial^2 N_k\langle r_k^{-1}\rangle}{\partial N_k^2}\right)^{\circ}(\Delta N_k)^2 + \cdots \tag{10.27}$$

We then define the energy

$$E_k^{\text{vs}} = -\frac{1}{\gamma_k^{\text{mol}}}Z_k^{\text{eff}}N_k\langle r_k^{-1}\rangle \tag{10.28}$$

of atom k in its valence state, in the current acceptation of this term, chosen so as to ensure the same interaction between the electrons of the atom as occurs when the atom is part of a molecule. The valence state is considered as being formed from a molecule by removing all the other atoms without allowing any electronic rearrangement in the atom of interest. All the interactions due to particles outside the volume occupied by this atom are "turned off" in E_k^{vs}. Now, taking the successive derivatives of E_k^{vs} evaluated for $N_k = N_k^{\circ}$, namely

$$\left(\frac{\partial E_k^{\text{vs}}}{\partial N_k}\right)^{\circ} = -\frac{Z_k^{\text{eff}}}{\gamma_k^{\text{mol}}}\left(\frac{\partial N_k\langle r_k^{-1}\rangle}{\partial N_k}\right)^{\circ}$$

$$\left(\frac{\partial^2 E_k^{\text{vs}}}{\partial N_k^2}\right)^{\circ} = -\frac{Z_k^{\text{eff}}}{\gamma_k^{\text{mol}}}\left(\frac{\partial^2 N_k\langle r_k^{-1}\rangle}{\partial N_k^2}\right)^{\circ},\cdots$$

it follows from Eq. (10.27) that

$$N_k\langle r_k^{-1}\rangle - N_k^{\circ}\langle r_k^{-1}\rangle^{\circ} = -\frac{\gamma_k^{\text{mol}}}{Z_k^{\text{eff}}}$$

$$\times \left[\left(\frac{\partial E_k^{\text{vs}}}{\partial N_k}\right)^{\circ}\Delta N_k + \frac{1}{2!}\left(\frac{\partial^2 E_k^{\text{vs}}}{\partial N_k^2}\right)^{\circ}(\Delta N_k)^2 + \cdots\right] \tag{10.29}$$

Up to here, it was convenient to carry out the calculations using the valence–electron populations N_k, N_l, and so on as variables. For the final step, however, it is more practical to express the results using *net* (i.e., nuclear minus electronic) charges, so that

$$\Delta q = -\Delta N \tag{10.30}$$

In this manner, one obtains from Eqs. (10.25), (10.29) and (10.30) that

$$\sum_{k<l} \varepsilon_{kl} = \sum_{k<l} \varepsilon_{kl}^{\circ} + \frac{1}{\gamma^{v}} \sum_{k} \left\{ \gamma_{k}^{\text{mol}} \left[\left(\frac{\partial E_{k}^{\text{vs}}}{\partial N_k} \right)^{\circ} \Delta q_k - \frac{1}{2!} \left(\frac{\partial^2 E_{k}^{\text{vs}}}{\partial N_k^2} \right)^{\circ} (\Delta q_k)^2 + \cdots \right] \right.$$

$$\left. - Z_k^{\text{eff}} \sum_{l} \langle r_{kl}^{-1} \rangle^{\circ} \Delta q_l \right\} + F \tag{10.31}$$

This equation contains the required information featuring the role of atomic charges in energy calculations. Molecular electroneutrality, namely, $\sum_k q_k = 0$, is ensured with the use of the appropriate Δq_k terms.

Now we want to write Eq. (10.31) in a more instructive fashion, in a form that highlights the properties of the individual bonds.

Final Energy Formulas

Let us define

$$a_{kl} = \frac{1}{v_k} \cdot \frac{\gamma_k^{\text{mol}}}{\gamma^{v}} \left[\left(\frac{\partial E_{k}^{\text{vs}}}{\partial N_k} \right)^{\circ} - \frac{1}{2!} \left(\frac{\partial^2 E_{k}^{\text{vs}}}{\partial N_k^2} \right)^{\circ} \Delta q_k + \cdots \right] - \frac{1}{\gamma^{v}} Z_l^{\text{eff}} \langle r_{kl}^{-1} \rangle^{\circ} \tag{10.32}$$

where v_k = number of atoms attached to k. Multiply a_{kl} by Δq_k and carry out the double sum $\sum_k \sum_l a_{kl} \Delta q_k$. Comparison with Eq. (10.31) shows that

$$\sum_{k<l} \varepsilon_{kl} = \sum_{k<l} \varepsilon_{kl}^{\circ} + \sum_{k} \sum_{l} a_{kl} \Delta q_k + F \tag{10.33}$$

or, for just the bond formed by atoms k and l

$$\varepsilon_{kl} = \varepsilon_{kl}^{\circ} + a_{kl} \Delta q_k + a_{lk} \Delta q_l + F_{kl} \tag{10.34}$$

$$F_{kl} = -\frac{1}{\gamma^{v}} Z_k^{\text{eff}} Z_l^{\text{eff}} \left[R_{kl}^{-1} - \left(R_{kl}^{-1} \right)^{\circ} - \left(\langle r_{kl}^{-1} \rangle - \langle r_{kl}^{-1} \rangle^{\circ} \right) \right]$$

$$- \frac{1}{\gamma^{v}} Z_k^{\text{eff}} q_l \left(\langle r_{kl}^{-1} \rangle - \langle r_{kl}^{-1} \rangle^{\circ} \right) \tag{10.35}$$

Equation (10.34) indicates how the intrinsic energy of a chemical bond linking atoms k and l depends on the electronic charges carried by the bond-forming atoms: ε_{kl}° is for a reference bond with net charges q_k° and q_l° at atoms k and l, respectively, whereas ε_{kl} corresponds to modified charges $q_k = q_k^{\circ} + \Delta q_k$ and $q_l = q_l^{\circ} + \Delta q_l$. F_{kl} follows from Eq. (10.26). These formulas are the basics of this theory. From here on, we focus on simplifications and application.

Convenient Simplifications

It turns to our advantage to consider ε_{kl}° and F_{kl} jointly. The idea is best explained by an example. Suppose that the CC and CH bond of ethane were selected as reference bonds, with energies ε_{CC}° and ε_{CH}°, respectively, and $F = 0$. New references are required for olefins, specifically tailored for $C(sp^2)$—$C(sp^3)$ and $C(sp^2)$—H bonds: the reference for $C(sp^2)$–$C(sp^3)$ bonds is deduced from that representing $C(sp^3)$—$C(sp^3)$ while $C(sp^2)$—H is derived from $C(sp^3)$—H by incorporating the appropriate parts of F into the new reference energies:

$$(\varepsilon_{kl}^{\circ})_{\text{original}} + F_{kl} = (\varepsilon_{kl}^{\circ})_{\text{modified}} \tag{10.36}$$

If we stipulate that ε_{kl}° includes F_{kl} by design, which makes sense because chemists know how to discriminate between different types of bonds, we can write the energy formula for a chemical bond in the following simple manner:

$$\varepsilon_{kl} = \varepsilon_{kl}^{\circ} + a_{kl}\Delta q_k + a_{lk}\Delta q_l \tag{10.37}$$

Another simplification facilitates things because it saves tedious work to satisfy charge normalization constraints. We write Eq. (10.37) as follows

$$\begin{aligned} \varepsilon_{kl} &= \varepsilon_{kl}^{\circ} + a_{kl}(q_k - q_k^{\circ}) + a_{lk}(q_l - q_l^{\circ}) \\ &= (\varepsilon_{kl}^{\circ} - a_{kl}q_k^{\circ} - a_{lk}q_l^{\circ}) + a_{kl}q_k + a_{lk}q_l \\ &= \varepsilon_{kl}^{\circ'} + a_{kl}q_k + a_{lk}q_l \end{aligned} \tag{10.38}$$

where

$$\varepsilon_{kl}^{\circ'} = \varepsilon_{kl}^{\circ} - a_{kl}q_k^{\circ} - a_{lk}q_l^{\circ} \tag{10.39}$$

defines a new reference, for a hypothetical kl bond between electroneutral atoms k and l. With this modification, featuring net charges $q_k, q_l, \ldots$ rather than their variations $\Delta q_k, \Delta q_l, \ldots$, our basic energy formula, Eq. (10.2), is written [44,108]

$$\Delta E_a^* = \sum_{k<l} \varepsilon_{kl}^{\circ'} + \sum_k \sum_l a_{kl}q_k - E_{\text{nb}} \tag{10.40}$$

While most convenient in computations, the $\varepsilon_{kl}^{\circ'}$ reference energy does not embrace an actual physical situation. The true physical reference bond, as it is found in an appropriately selected reference molecule, is ε_{kl}°, with reference atomic charges q_k° and q_l° at the bond-forming atoms k and l, respectively: ε_{kl}° is amenable to *direct* calculations; $\varepsilon_{kl}^{\circ'}$ is not.

Finally, regarding the a_{kl} parameters, Eq. (10.32) is our working formula, but in most cases one can do with the simpler form

$$a_{kl} = \frac{1}{v_k}\left(\frac{\partial E_k^{\text{vs}}}{\partial N_k}\right)^{\circ} - \frac{3}{7}\cdot\frac{Z_l^{\text{eff}}}{R_{kl}} \tag{10.41}$$

where the second-order derivatives are omitted, with $\gamma^{\text{v}} \simeq \gamma_k^{\text{mol}}$ set to $\frac{7}{3}$ for atoms other than hydrogen.

The quantity $\sum_l a_{kl}\Delta q_k$ reduced to essentials by means of Eq. (10.41) reveals its simple physical content. When the total energy of atom k in the molecule is varied by $(\partial E_k^{vs}/\partial N_k)^{\circ}\Delta q_k$, this Δq_k concurrently modifies the total energy of atom k by $-\frac{3}{7}\sum_l (Z_l^{eff} \times \Delta q_k)/R_{kl}$.

This concludes the presentation of bond energy formulas.

10.5 BASIC THEORETICAL PARAMETERS

This section describes the evaluation of the basic parameters γ_k^{at}, γ_k^{mol}, γ^v, $(\partial E_k^{vs}/\partial N_k)^{\circ}$, and $(\partial^2 E_k^{vs}/\partial N_k^2)^{\circ}$, ready for use in Eqs. (10.12) and (10.32).

The γ Parameters

The γ parameters of the isolated atoms γ_k^{at} are readily obtained from Eqs. (3.20) or (3.21). Selected SCF results[4] are indicated in Table 10.2. For hydrogen, of course, $\gamma = 2$ because of the virial theorem.

For atoms in a molecule, we use Eq. (4.10) and rewrite it as follows

$$E_k = \frac{1}{\gamma_k^{mol}}\left(V_{ne,k} + Z_k \sum_{l \neq k} \frac{Z_l}{R_{kl}}\right) \tag{10.42}$$

with the help of Eq. (4.2), where $V_{ne,k} = -Z_k \int [\rho(\mathbf{r})/(|\mathbf{r} - \mathbf{R}_k|)]d\mathbf{r}$. These γ values are obtained [90,213] from the SCF potentials at the individual nuclei $k, l, \ldots$ and from their fit with total energies using Eqs. (4.12) and (10.42), assuming a constant γ_k^{mol} for each atomic species. The latter point merits attention because, as seen in Eq. (10.42), $1/\gamma_k^{mol}$ multiplies a potential energy consisting of two distinct contributions, namely, a nuclear–electronic and a nuclear–nuclear part. Now, one can imagine to use two independent multipliers, $1/\gamma_k^{el}$ and $1/\gamma_k^{nucl}$, and write

$$E_k = \frac{1}{\gamma_k^{el}}V_{ne,k} + \frac{1}{\gamma_k^{nucl}}Z_k \sum_{l \neq k}\frac{Z_l}{R_{kl}} \tag{10.43}$$

by letting

$$\frac{1}{\gamma_k^{mol}} = \left(\frac{1}{\gamma_k^{el}}V_{ne,k} + \frac{1}{\gamma_k^{nucl}}Z_k \sum_{l \neq k}\frac{Z_l}{R_{kl}}\right)\frac{1}{V_k} \tag{10.44}$$

that is, by taking $1/\gamma_k^{mol}$ as the weighted average of $1/\gamma_k^{el}$ (for the nuclear–electronic part) and $1/\gamma_k^{nucl}$ (for the nuclear–nuclear part). Detailed SCF computations indicate that $\gamma_k^{el} = \gamma_k^{nucl}$, at least within the precision permitted by this type of analysis. This result settles an important question regarding the constancy of the γ_k^{mol} parameter introduced in the defining equation (4.10). Indeed, if it were $\gamma_k^{el} \neq \gamma_k^{nucl}$, then

---

[4]The atomic wavefunctions are from Ref. 212. Additional results are reported in Refs. 27 and 213.

γ_k^{mol} would change to some extent from molecule to molecule, depending on the relative weights of $V_{\text{ne},k}$ and $Z_k \sum_{l \neq k} Z_l / R_{kl}$ in Eq. (10.44). The results are included in Table 10.2.

In closing, let us examine the γ values for molecules using Eq. (4.11). It is clear that the average $1/\gamma$ depends on the γ_k^{mol} values of Table 10.2 and on the weights of the potential energies V_k of the individual atoms. Now, the weights of the heavy atoms, of the order of -88 au for carbon and -175 au for oxygen, for example, are considerably larger than that of hydrogen, ~ -1 au. As a consequence, the final result for γ is in most cases close to $\frac{7}{3}$. This is a well-known fact: SCF calculations made for a great variety of organic molecules indicate that $\gamma \simeq \frac{7}{3}$ within approximately $\pm 1\%$. We use this approximation for γ^v, keeping in mind that it affects only a small portion of the total energy of atomization.

Calculation of $(\partial E_k^{\text{vs}}/\partial N_k)^\circ$ and $(\partial^2 E_k^{\text{vs}}/\partial N_k^2)^\circ$

We now direct our attention to the calculation of the a_{kl} parameters. The first and second derivatives, $(\partial E_k^{\text{vs}}/\partial N_k)^\circ$ and $(\partial^2 E_k^{\text{vs}}/\partial N_k^2)^\circ$, are most conveniently obtained from SCF–Xα theory [174], which offers the advantage of permitting calculations for any desired integer or fractional electron population. It is, indeed, important to account for the fact that these derivatives depend on N_k. The difficulty is that calculations of this sort cannot be performed directly for atoms that are actually part of a molecule. So one resorts to model free-atom calculations to mimic the behavior of atoms that are in a molecule but do not experience interactions with the other atoms in the host molecule (Table 10.3).

For hydrogen, $\alpha = 0.686$ is appropriate for a partially negative atom, like that of ethane, $q_{\text{H}} = -11.7$ me, and reproduces its electron affinity.

TABLE 10.3. Selected First and Second Energy Derivatives (au)

Atom	Orbital	Population	$(\partial E_k/\partial N_k)^\circ$	$(\partial^2 E_k/\partial N_k^2)^\circ$
H	$1s$	1.0117	-0.195	0.40
	$1s^a$	1.000	-0.200	—
C	$2s$	1.44	-0.735	0.45
	$2p$	1.97	-0.200	—
	σ	3.00	-0.375	—
C(C_2H_4)	π	1.00	-0.246	—
C(C_6H_6)	π	1.00	-0.262	—
N	$2p$	3	-0.28	—
O (ethers)	$2p$	3.990	-0.333	0.50
O (alcohols)	—	—	-0.80	—

[a] For a_{HO} we have used $(\partial E/\partial N)_{\text{H}}^\circ = -0.200$ au, corresponding to an approximately null hydrogen net charge. This derivative follows from spin-restricted Xα calculations (with $\alpha = 0.686$) indicating that $E_{\text{H}} = -0.72650N + 0.324525N^2 - 0.040833N^3$ hartree, with $N =$ number of electrons. For the ether oxygen atom we have used $(\partial E/\partial N)_{\text{O}}^\circ = -0.333$ au, but -0.80 for the hydroxyl oxygen in the evaluation of a_{OH} and a_{OC} this value was suggested by accurate energy calculations of alcohols.

Optimized SCF computations indicate that for the carbon atoms of saturated hydrocarbons any gain in electronic charge, with respect to the ethane carbon, occurs at the $2s$ level [44]. Further, $(9s\,5p|6s) \rightarrow [5s\,3p|3s]$ calculations of methane and ethane, using Dunning's exponents [85] and optimum contraction vectors, reveal C($2s$) populations of 1.42–1.46 e. Finally, SCF–Xα computations give $(\partial E_k/\partial N_k)$ values of -20.29, -19.87, and -19.26 eV for C($2s$) populations of 1.40, 1.45, and 1.50 e, respectively, by using the $\alpha = 0.75928$ value recommended by Schwarz [214]. These results suggest that the appropriate $(\partial E_k/\partial N_k)$ derivative can be reasonably estimated at -0.735 au $(-20\,\text{eV})$ for sp^3 carbon atoms. The results for sp^2 carbons [129] follow from similar SCF–Xα calculations. The result for O($2p$) was deduced using the "frozen core" α value given by Schwarz, $\alpha = 0.74447$. The result indicated for nitrogen is from Ref. 215.

10.6 SATURATED MOLECULES

Any application of Eqs. (10.37)–(10.40) requires a solid knowledge of the appropriate set of reference bond energies ε_{kl}°, of the bond energy parameters a_{kl} and, finally, of the appropriate atomic charges.

The data of Tables 10.2 and 10.3 give access to the a_{kl} parameters listed in Table 10.4. Concerning the $\varepsilon_{kl}^{\circ'}$ energies, we are henceforth in a position to benefit from detailed solutions obtained for the alkane molecules, but for the details regarding nitrogen- and oxygen-containing molecules we must refer to Chapters 15 and 16, respectively.

With the second-order derivatives being usually omitted in Eq. (10.32) owing to the smallness of the Δq_k values, the a_{kl} terms can be treated as constants.

Calculation of Reference Bond Energies [Eq. (10.39)]. The a_{kl} parameters indicated in Table 10.4 are ready for use in the bond energy formula, Eq. (10.39). The following examples, in part based on detailed results given in Chapters 15 and 16 for nitrogen- and oxygen-containing molecules, illustrate the procedure and report the input data.

TABLE 10.4. Reference Bond Energies (kcal/mol) and a_{kl} Parameters

Bond	R_{kl} (Å)	Occurrence	$\varepsilon_{kl}^{\circ'}$	a_{kl} (kcal mol^{-1} me^{-1})^a	
C—C	1.531	Alkanes	103.891	$a_{CC} = -0.488$	—
N—N	1.446	Hydrazine	29.25	$a_{NN} = -0.551$	—
C—N	1.46	Amines	75.56	$a_{CN} = -0.603$	$a_{NC} = -0.448$
C—O	1.43	Ethers	104.635	$a_{CO} = -0.712$	$a_{OC} = -0.501$
C—O	1.43	Alcohols	104.635	$a_{CO} = -0.712$	$a_{OC} = -0.649$
C—H	1.08	Alkanes	108.081	$a_{CH} = -0.247$	$a_{HC} = -0.632$
N—H	1.032	Amines	101.36	$a_{NH} = -0.197$	$a_{HN} = -0.794$
O—H	0.957	Alcohols	115.678	$a_{OH} = -0.400$	$a_{HO} = -1.000$

^{a}Conversion factors: 1 hartree = 627.51 kcal/mol; 1 bohr = 0.52917 Å.

For the CC bond in ethane (see below), we find $\varepsilon^\circ_{CC} = 69.633$ kcal/mol, with atomic charges of 35.1 me at the carbon atoms; for its CH bonds, we find $\varepsilon^\circ_{CH} = 106.806$ kcal/mol, with $q_C = 35.1$ and $q_H = -11.7$ me. Similarly, we have $\varepsilon^\circ_{CO} = 79.78$ kcal/mol for the CO bond in diethylether, with $q^\circ_C = 31.26$ and $q^\circ_O = 5.18$ me for the bond-forming atoms. For the CN bond in methylamine, we have $\varepsilon^\circ_{CN} = 60.44$ kcal/mol, with $q^\circ_C = 31.77$ and $q^\circ_N = -9.00$ me, and $\varepsilon^\circ_{NH} = 100.99$ kcal/mol for $q^\circ_H = 0.10$ me [139]. This leads to $\varepsilon_{NH} = 101.36 - 0.197q_N - 0.794q_H$. Next, assuming $\varepsilon_{NN} = 36.30$ kcal/mol ($= \varepsilon^\circ_{NN}$) for hydrazine (deduced from $\Delta E^*_a = 436.62$ kcal/mol, with $\Delta H^\circ_f(\text{gas}) = 22.79$ kcal/mol [216] and $\text{ZPE} + H_T - H_0 = 34.14$ kcal/mol [217]), we get $q_N(\text{hydrazine}) = -6.40$ me and thus $\varepsilon^{\circ\prime}_{NN} = 29.25$ kcal/mol. Finally, taking water as reference molecule, calculations indicate $\varepsilon^{\circ\prime}_{OH} = 115.678$ kcal/mol (Chapter 16.2).

Saturated Hydrocarbons

To help develop a familiarity with Eq. (10.37), we examine a general formula for saturated hydrocarbons, $C_nH_{2n+2-2m}$, containing n carbon atoms and m chair or boat six-membered cycles. These alkanes contain $(n-1+m)$ CC bonds and $(2n + 2 - 2m)$ CH bonds; hence

$$\sum_{k<l} \varepsilon^\circ_{kl} = (n-1+m)\varepsilon^\circ_{CC} + (2n+2-2m)\varepsilon^\circ_{CH} \tag{10.45}$$

Each carbon forms N_{CC} bonds with other carbon atoms and $4 - N_{CC}$ bonds with hydrogen atoms. The part of $\sum_l a_{kl}\Delta q_k$ due to that C atom is $N_{CC}a_{CC}\Delta q_C + (4 - N_{CC})a_{CH}\Delta q_C$. On the other hand, each H atom contributes $a_{HC}\Delta q_H$. Summation over all C and H atoms gives

$$\sum_k \sum_l a_{kl}\Delta q_k = (a_{CC} - a_{CH}) \sum N_{CC}\Delta q_C + 4a_{CH} \sum \Delta q_C$$

$$+ a_{HC} \sum \Delta q_H \tag{10.46}$$

Using these results in Eq. (10.2), one obtains, with $A_1 = a_{CC} - a_{CH}$ and $A_2 = 4a_{CH} - a_{HC}$, the following equation:

$$\Delta E^*_a = (n - 1 + m)\,\varepsilon^\circ_{CC} + (2n + 2 - 2m)\,\varepsilon^\circ_{CH} + A_1 \sum N_{CC}\Delta q_C$$

$$+ A_2 \sum \Delta q_C + (n - 2 + 2m)\,a_{HC}q^\circ_H - E_{nb} \tag{10.47}$$

Equation (10.47) lends itself to three instructive tests.

The first test regards the atomic charges. We write Eq. (10.47) for a representative selection of molecules using explicitly the net charges given by Eqs. (8.7) and (8.8), that is (where superscript "Mull" denotes Mulliken)

$$q_C = q_C^{Mull} + N_{CH} \times p$$
$$q_H = q_H^{Mull} - p$$

and determine the unknown p by a least-square analysis with the help of experimental ΔE_a^* data. For fully optimized STO-3G charges, one obtains [108] $p = (30.3 \pm 0.3) \times 10^{-3}$ e. Therefore, from what we have learned in Chapters 6–8, we can use the ^{13}C NMR shifts for deriving the required Δq_C charges and write $A_1 \sum N_{CC} \delta_C$ instead of $A_1 \sum N_{CC} \Delta q_C$ and $A_2 \sum \delta_C$ instead of $A_2 \sum \Delta q_C$. The "chemical definition" of atomic charges offered in Chapter 5—a clear case of chemical prejudice that led to Eqs. (5.10) and (6.7)—is unmistakably confirmed.

The second test concerns the parameters A_1 and A_2. They are readily obtained from least-square fittings of experimental ΔE_a^* data with Eq. (10.47). The empirical ratio $A_2/A_1 = 1.486$ thus determined equals that anticipated from the theoretical values, $a_{CC} = -0.777$, $a_{CH} = -0.394$ and $a_{HC} = -1.007$ au. So we have $A_1 = -0.383$ and $A_2 = -0.569$ au, which are conveniently expressed in kcal mol^{-1} ppm^{-1} units as $A_1 = 0.383(627, 51/237.1) q_C^\circ$ and $A_2 = 0.569(627.51/237.1) q_C^\circ$.

The final test uses these A_1 and A_2 parameters in applications of Eq. (10.47). Solving it for ethane, we find for its bonded part that $\Delta E_{a,bonded}^{*C_2H_6} = \varepsilon_{CC}^\circ + 6\varepsilon_{CH}^\circ$. Similarly, we get for methane that $\Delta E_{a,bonded}^{*CH_4} = 4\varepsilon_{CH}^\circ + A_2 \delta_C^{CH_4} - a_{HC} q_H^\circ$, where $\delta_C^{CH_4}$ ($= -8$ ppm) is the NMR shift of the methane carbon in ppm from ethane. Now we rearrange Eq. (10.47):

$$2n\varepsilon_{CC}^\circ = \Delta E_a^* + E_{nb} + n\left(\Delta E_{a,bonded}^{*C_2H_6} - 2\Delta E_{a,bonded}^{*CH_4} + 2A_2 \delta_C^{CH_4} + a_{HC} q_C^\circ\right)$$
$$+ (1 - m)\left(\Delta E_{a,bonded}^{*C_2H_6} - 2\Delta E_{a,bonded}^{*CH_4} + 2A_2 \delta_C^{CH_4}\right)$$
$$- A_1 \sum N_{CC} \delta_C - A_2 \sum \delta_C \tag{10.48}$$

and take advantage of the fact that ε_{CC}° is constant by definition. Note the presence of q_C° in the right-hand side of (10.48), namely, in A_1 and A_2. Application of Eq. (10.48) to a selection of molecules indicates that the constraint of a constant ε_{CC}° is satisfied for $q_C^\circ \simeq 0.035$ e, with $\varepsilon_{CC}^\circ \simeq 69.7$ kcal/mol [108]. Our optimum choice, $q_C^\circ = 35.1$ me, yields $A_1 = 0.0356$ and $A_2 = 0.0529$ kcal/mol^{-1} ppm^{-1} with $\varepsilon_{CC}^\circ = 69.633$ kcal/mol.

Empirical Feigned Bond Additivity. Let us put our results in perspective with respect to brute-force empirical fits intended to define "best possible" sets of transferable bond energies, $\varepsilon_{kl}^{empir.}$. Remembering that for ethane $q_C^\circ + 3q_H^\circ = 0$, we can write the following identity:

$$(n - 2 + 2m) q_H^\circ = -\frac{1}{2}(n - 1 + m) q_C^\circ - \frac{1}{4}(2n + 2 - 2m) q_H^\circ$$

Combining this identity with Eq. (10.47), it appears that

$$\sum_{k<l} \varepsilon_{kl} = \mathcal{A} + (n - 1 + m)(\varepsilon_{CC}^\circ - \frac{1}{2}a_{HC}q_C^\circ) + (2n + 2 - 2m)(\varepsilon_{CH}^\circ - \frac{1}{4}a_{HC}q_H^\circ)$$

with $\mathcal{A} = A_1 \sum N_{CC}\Delta q_C + A_2 \sum \Delta q_C$.

A multiple regression analysis that uses experimental $\Delta E_a^* \approx \sum_{k<l} \varepsilon_{kl}$ values and simply counts the $(n - 1 + m)$ CC bonds and the $(2n + 2 - 2m)$ CH bonds would view $\varepsilon_{CC}^\circ - \frac{1}{2}a_{HC}q_C^\circ = 80.7$ kcal/mol as *the* CC bond energy, $\varepsilon_{CC}^{empir.}$, and $\varepsilon_{CH}^\circ - \frac{1}{4}a_{HC}q_H^\circ = 105.0$ kcal/mol as *the* CH bond energy, $\varepsilon_{CH}^{empir.}$, in lieu of $\varepsilon_{CC}^\circ = 69.633$ and $\varepsilon_{CH}^\circ = 106.806$ kcal/mol. Charge normalization terms are inadvertently treated as part of bond energy! Both $\varepsilon_{CC}^{empir.}$ and $\varepsilon_{CH}^{empir.}$ are metaphors, not physical entities. In the final count, $\mathcal{A}$, which is usually interpreted as being due to "steric effects," is simply a function of local charge variations.

Formula for Alkanes Including Nonbonded Interactions. Consider Eq. (10.47) and rewrite it as follows:

$$\Delta E_a^* = (1 - m)(\varepsilon_{CC}^\circ + 6\varepsilon_{CH}^\circ)$$
$$+ (n - 2 + 2m)(\varepsilon_{CC}^\circ + 6\varepsilon_{CH}^\circ - 4\varepsilon_{CH}^\circ + a_{HC}q_H^\circ - A_2\delta_C^{CH_4})$$
$$+ (n - 2 + 2m)A_2\delta_C^{CH_4} + A_1 \sum N_{CC}\delta_C + A_2 \sum \delta_C - E_{nb}$$

Using the quantities $\Delta E_{a,bond.}^{*C_2H_6}$ and $\Delta E_{a,bond.}^{*CH_4}$ defined above, we get

$$\Delta E_a^* = (1 - m)\Delta E_{a,bond.}^{*C_2H_6} + (n - 2 + 2m)\left(\Delta E_{a,bond.}^{*C_2H_6} - \Delta E_{a,bond.}^{*CH_4}\right)$$
$$+ (n - 2 + 2m)A_2\delta_C^{CH_4} + A_1 \sum N_{CC}\delta_C + A_2 \sum \delta_C - E_{nb}$$

Finally, using the approximation (10.8) for E_{nb} and applying Eq. (10.2), we deduce that

$$\Delta E_a^* \simeq (1 - m)\Delta E_a^{*C_2H_6} + (n - 2 + 2m)\left(\Delta E_a^{*C_2H_6} - \Delta E_a^{*CH_4}\right)$$
$$+ (n - 2 + 2m)A_2\delta_C^{CH_4} + A_1 \sum N_{CC}\delta_C + A_2 \sum \delta_C \qquad (10.49)$$

Now we know that nonbonded energies differ somewhat from case to case in comparisons between structural isomers. On the other hand, the $\sum N_{CC}\delta_C$ and $\sum \delta_C$ terms are also structure-dependent. For these reasons, a minor readjustment of the parameters A_1 and A_2 succeeds in compensating part of the error introduced by assuming exact additivity for the nonbonded Coulomb contributions. The recommended values are [190], in $kcal\,mol^{-1}\,ppm^{-1}$ units, $A_1 = 0.03244$ and $A_2 = 0.05728$. With $\Delta E_a^{*C_2H_6} = 710.54$, $\Delta E_a^{*CH_4} = 419.27$ kcal/mol and ^{13}C shifts in ppm from ethane, Eq. (10.49) becomes, in kcal/mol units

$$\Delta E_a^* \simeq 710.54(1 - m) + 290.812(n - 2 + 2m)$$
$$+ 0.03244 \sum N_{CC}\delta_C + 0.05728 \sum \delta_C \qquad (10.50)$$

This handy energy formula requires only the ^{13}C NMR spectra of the molecules under scrutiny. A comparison made for a group of 19 molecules indicated a root-mean-square deviation of 0.25 kcal/mol relative to experimental data, whereas the rms deviation amounts to 0.21 kcal/mol for calculations made with the theoretical A_1 and A_2 parameters, with nonbonded energies deduced directly from Eq. (10.3).

This concludes the derivation of our bond energy formula and the presentation of simple examples pertaining to saturated systems, namely, the alkanes, including their numerical parameterization.

Unsaturated hydrocarbons are our next target.

CHAPTER 11

THE CHEMICAL BOND: THEORY (II)

Chapter 10 has taught us that this theory is that of a simple formula for bond energies, namely, Eq. (10.37)

$$\varepsilon_{kl} = \varepsilon_{kl}^{\circ} + a_{kl}\Delta q_k + a_{lk}\Delta q_l$$

or its companion, Eq. (10.38), featuring net charges q_k and q_l rather than their variations, Δq_k and Δq_l. Formulas are given for ε_{kl}° and a_{kl}, as well as selected numerical values, namely, all those required for the alkanes, as they were deduced from the appropriate theoretical data.

Tests are convincing both because they were entirely based on these theoretical parameters and because they were conducted with atomic charges that are supported by strong arguments suggested by a number of independent sources, including SDCI charge analyses, and satisfy a most accurate correlation with ^{13}C NMR shifts [Eq. (6.8)]. Moreover, they are uncomplicated: the calculation of bond energies and hence of atomization energies, $\Delta E_a^* = \sum_{k<l} \varepsilon_{kl} - E_{nb}$, are straightforward, although most of the work involved is in the evaluation of the very small nonbonded part, E_{nb}.

A considerable simplification is achieved with a general formula [Eq. (10.50)], which incorporates these nonbonded contributions in an approximate manner. The unsigned average deviation with respect to the experimental results of seventyone saturated hydrocarbons, 0.19 kcal/mol, certainly supports this simplified formula (see Part III).

Atomic Charges, Bond Properties, and Molecular Energies, by Sándor Fliszár
Copyright © 2009 John Wiley & Sons, Inc.

In order to extend the applicability of the same simple methods to olefinic, polyenic, and aromatic unsaturated hydrocarbons, additional parameters are required: those for the bonds involving sp^2 carbons. One must thus consider

- The modification of reference charges that involve readjustments of reference bond energies, evaluated by the function $a_{kl}\Delta q_k$
- The description of σ and π charge centroids in unsaturated systems, and their role in energy calculations, evaluated by means of the function F_{kl}, Eq. (10.35)
- The inverse variations of σ and π electron populations in aromatic and olefinic sp^2 systems and their incidence in the calculation of the appropriate a_{kl} parameters
- Conjugation and its inclusion in the relevant bond energies ε_{kl}°

Accordingly, 10 new reference bond energies must be defined. This is done with respect to the well-known bonds of ethane. In addition, seven additional a_{kl} parameters are to be evaluated by means of our standard formulas, Eqs. (10.41) and (11.16), with proper consideration of σ- and π-electron populations. All these transformations, including those dictated by changes of internuclear distances, are straightforward but require attention.

The nice thing worth mentioning here is that they need be done only once. The appropriate transformations are presented in detail in order to illustrate in what manner they are tightly interrelated. But then, the newly defined ε_{kl}° and a_{kl} parameters serve most simply in our basic formulas (10.37) or (10.38), whose use is quick, easy, and accurate.

11.1 VALENCE ATOMIC ORBITAL CENTROIDS

The description presented so far has dealt primarily with the *numbers* of electrons associated with the individual atoms in a molecule. Now we examine the *shape* of these electron populations. The electron densities ρ and electron populations N are those of the valence region.

The external electrostatic potential at atom k is

$$\mathcal{V}_k = -\sum_l \left[\int \frac{\rho_l(\mathbf{r})}{|\mathbf{r} - \mathbf{R}_k|} \, d\mathbf{r} - Z_l^{\mathrm{eff}} R_{kl}^{-1} \right]$$

where $\rho_l(\mathbf{r})$ is the density of the N_l valence electrons assigned to atom $l \neq k$, and Z_l^{eff} is its effective nuclear charge, at a distance R_{kl} from k.

Using Eq. (10.19), we write

$$\mathcal{V}_k = -\sum_l (N_l \langle r_{kl}^{-1} \rangle - Z_l^{\mathrm{eff}} R_{kl}^{-1}) \tag{11.1}$$

Let us also introduce the corresponding expression obtained when the electron populations are kept at their proper values in the molecule under study, but the inverse-distance terms are replaced by their values (indicated by the superscript "○") of a model reference molecule:

$$V_k^{nr} = -\sum_l [N_l \langle r_{kl}^{-1} \rangle^\circ - Z_l^{eff}(R_{kl}^{-1})^\circ] \tag{11.2}$$

The difference $V_k - V_k^{nr}$ describes a change in electrostatic potential at atom k. The corresponding change in potential energy, summed over all atoms, is

$$f = \sum_k Z_k^{eff}(V_k - V_k^{nr}) \tag{11.3}$$

This is the quantity we shall use for our discussion. It is a convenient way of gaining insight into the function F [Eq. (10.26)] because

$$f = -\gamma^v F \tag{11.4}$$

Extensive numerical calculations indicate that f is (1) negligible (say, <0.3 kcal/mol) for saturated hydrocarbons but (2) significant for olefinic molecules (e.g., ~ 40 kcal/mol for tetramethylethylene). The condition that f should vanish can be satisfied either because the various atomic contributions to f cancel or because the individual terms in the summation over k vanish. Since it seems unlikely that cancellation of terms associated to different atoms would take place systematically in a large number of molecules, we shall assume that, to a good approximation

$$V_k \simeq V_k^{nr} \tag{11.5}$$

for any atom of any saturated hydrocarbon.

Now, the most direct interpretation of Eq. (11.5) follows from the observation, suggested by Eqs. (11.1) and (11.2), that f is essentially a "relaxation" term. In fact, $V_k - V_k^{nr}$ represents the difference between the electrostatic potential at the kth nucleus in the given molecule and the potential that the same nucleus would feel if the atomic orbitals and the equilibrium distances remained the same as in the reference molecule in spite of the change in electron populations.

With this picture in mind, Eq. (11.5) reads:

Whenever the atoms under consideration in a given molecule are in the same valence states as in the reference molecule, the relaxation process is such that the potential created by the other atoms at the kth nucleus is the same as would be predicted by leaving the pertinent internuclear distances and the shapes of atomic electron densities as they are in the reference molecule, with the electron populations changed as required by the new situation.

This may mean that changes in the nuclear positions and in the centroids of the atomic orbitals always take place so as to leave the ratio between the expectation value $\langle r_{kl}^{-1} \rangle$ and R_{kl}^{-1} the same as in the reference molecule or, at least, that the changes are insignificant. It is impossible to decide which alternative applies from a study of the paraffins, since their bond distances and atomic orbitals are expected to be practically constant.

A simple way of satisfying Eq. (11.5) consists in assuming that

$$\langle r_{kl}^{-1} \rangle \simeq R_{kl}^{-1} \tag{11.6}$$

when all bonds are σ bonds at their equilibrium geometries, as in the alkanes. This simplifying hypothesis is supported by examination of SCF potentials at the nuclei, showing that Eq. (11.6) holds to within $\sim 5\%$, for both CH and CC bonds. It is especially useful here because it permits us to isolate the relevant conceptual points in a straightforward manner within the simple framework of the point-charge approximation. Defining a density $\rho_{\mu l}$ and an electron population $N_{\mu l}$ associated with each atomic orbital μ of atom l, so that

$$\int \frac{\rho_{\mu l}(\mathbf{r})}{|\mathbf{r} - \mathbf{R}_k|} \, d\mathbf{r} = N_{\mu l} \langle r_{kl}^{-1} \rangle \tag{11.7}$$

we can write[1]

$$\sum_{\mu} \frac{N_{\mu l}}{N_l} \langle r_{\mu kl}^{-1} \rangle = 1/R_{kl} \tag{11.8}$$

Equation (11.8) reads: "The average of the expectation values of $|\mathbf{r} - \mathbf{R}_k|^{-1}$ for the various valence AOs of atom l, weighted by the rations of the orbital populations to the total atomic population of atom l equals the inverse of the $k - l$ distance."

For all their their simplicity, Eqs. (11.7) and (11.8) cannot be tested numerically by direct calculation. The reason is linked to the difficulty of partitioning the total electron density into atomic contributions, in spite of an important conceptual step forward due to Parr [219]. A practical step in the same direction is in the construction of suitable in situ valence atomic orbitals (VAO) from accurate ab initio computations [143], as advocated long ago by Mulliken [220] and discussed by Del Re [221]. As will be seen, such in situ VAOs do provide useful information, but they are of no help in solving the additional problem of assigning suitable populations to the orbitals and of dividing overlap populations into atomic contributions. In view of this situation, we take Eqs. (11.5) and (11.8) as statements whose validity rests on experimental evidence, at least for saturated hydrocarbons.

Check Test: The Olefins

In paraffins, the bonds formed by carbon are either CH or CC single bonds; of course, the electron density around C is not exactly spherical, but special multipole

---

[1]Equation (11.7) implies that orbital products can well be approximated in the average by some Mulliken-type expansion [206,218].

contributions appear to vanish to the extent that one can assume that the CH electron density adjusts to that of CC, and conversely, so that the electrostatic effects are adequately described by a point charge located at C. Now suppose that one bond to hydrogen is removed and replaced by a π bond of the same atom to another carbon atom. Changes in geometry and hybridization will, of course, occur, but, at least to first order, the resulting π atomic orbital will not be able to participate in a possible mutual readjustment of the AO centroids, due to its different symmetry. Thus we have to expect that condition (11.8) also applies to olefins, with the restriction that it concerns only σ orbitals. In other words, ethylenic carbon atoms will obey a "principle" of σ-bond electrostatic balancing [222]:

$$N_l\langle r_{kl}^{-1}\rangle = \int \frac{\rho_l(\mathbf{r})}{|\mathbf{r} - \mathbf{R}_k|}\, d\mathbf{r} = \sum_\mu \int \frac{\rho_{\mu l}(\mathbf{r})}{|\mathbf{r} - \mathbf{R}_k|}\, d\mathbf{r}$$

$$= \frac{N_l - N_{\pi l}}{R_{kl}} + N_{\pi l}\langle r_{\pi kl}^{-1}\rangle \qquad (11.9)$$

Let us now consider $\langle r_{\pi kl}^{-1}\rangle$. A π atomic orbital is essentially a pure p orbital. If there is any polarization (as will be discussed below), this will involve a very small displacement $\Delta r_{\pi l}$ of the centroid of the orbital. Then a monopole approximation of the π term of Eq. (11.9) suffices, giving

$$\langle r_{kl}^{-1}\rangle = \frac{1 - N_{\pi l}/N_l}{R_{kl}} + \frac{N_{\pi l}/N_l}{|\mathbf{r}_l + \Delta\mathbf{r}_{\pi l} - \mathbf{R}_k|} \qquad (11.10)$$

If this simple expression is sufficient to account for the discrepancies observed in calculations of olefins based on alkane reference bonds [i.e., calculations using Eq. (10.33) with $F = 0$], we can claim that the "principle" (11.8) is, indeed, satisfied by ethylenic hydrocarbons.

Turning now to direct theoretical evaluations, we consider $|\Delta\mathbf{r}_{\pi l}|$ as the displacement (on the C=C axis) of the centroid of the π orbital with respect to the center l. Of course, such a displacement can differ from zero only if some hybridization is allowed, which, in the case of a π orbital, must consist in admixture of the suitable $d\pi$ orbital. The hybrid in question was determined [222] from 4-31G calculations with d polarization functions for carbon and optimization of all scale factors, followed by a calculation of in situ valence orbitals, and of their characteristics, according to Del Re and Barbier [143]. The inward shifts (on the C=C axis) of the π orbital centroids are close to 0.03 Å (Table 11.1).

Applications of Eq. (11.10) are straightforward, using

$$|\mathbf{r}_l + \Delta\mathbf{r}_{\pi l} - \mathbf{R}_k| = \sqrt{R^2 + (\Delta r_{\pi l})^2 - 2R\Delta r_{\pi l}\cos\varphi} \qquad (11.11)$$

where R is the length of the C(sp^2)—k bond ($k = $ H or C(sp^3)) and φ is the angle it forms with the axis of the carbon–carbon double bond. Experimental geometries and

TABLE 11.1. Inward Shifts, $\Delta r_{\pi l}$ on the C=C Axis, of π-Orbital Centroids

Molecule	Bond kl	Δr (Å)
Ethylene	H—C(sp^2)	0.0292
Propene	H—C$_1$(sp^2)	0.0281
	H—C$_2$(sp^2)	0.0299
Tetrametylethylene	C(sp^3)—C(sp^2)	0.0280

the ethylene reference populations $N_{\pi l} = 1$ and $N_l = 3.9923$ e were used [222] to solve (11.10) with the help of (11.11). Thus we deduce F from Eq. (10.35), given here for $\gamma^v = \frac{7}{3}$:

$$F_{kl} = -\frac{3}{7} Z_k^{\text{eff}} Z_l^{\text{eff}} \left[R_{kl}^{-1} - (R_{kl}^{-1})^\circ - \left(\langle r_{kl}^{-1} \rangle - \langle r_{kl}^{-1} \rangle^\circ \right) \right]$$

$$- \frac{3}{7} Z_k^{\text{eff}} q_l \left(\langle r_{kl}^{-1} \rangle - \langle r_{kl}^{-1} \rangle^\circ \right). \tag{11.12}$$

For all their simplicity, the ideas embodied in Eq. (11.10) offer a realistic interpretation explaining the essence of the modifications taking place in CC and CH bonds due to the replacement of an sp^3-hybridized carbon by an sp^2 carbon atom.

Despite the marked differences in both geometric parameters and the SCF Δr values between the molecules involved in this comparison, there are striking regularities: the F value calculated for propene is, for all practical reasons, $\frac{3}{4}$th that of ethylene (which takes care of 3 CH bonds) plus $\frac{1}{4}$th that of tetramethylethylene (for the CC bond). Capitalizing on this idea, we may well consider transferable bond contributions modeled after Eq. (11.12) and use them to generate new reference bond energies satisfying Eq. (10.36).

This considerably simplified approach works extremely well in molecular calculations. The bottom line is that we can safely proceed with Eq. (10.37) as long as we use the appropriate reference bond energies ε_{kl}°, because these reference energies enjoy a high degree of transferability.

Geometry Changes and Displacements of Charge Centroids

The problems to be solved are best illustrated by a typical example. Take ε_1°, the carbon–carbon bond energy in ethane with $R_{CC}^\circ = 1.531$Å; calculate the energy of a C(sp^3)—C(sp^2) bond like that found in olefins. The latter is for reference charges, which are 35.1 me for the sp^3 carbon (as in ethane) and 7.7 me for the sp^2 carbon (as in ethylene). This transformation is schematically represented in Fig. 11.1.

This transformation involves (1) a change of charge, (2) a change of the bond length, and (3) the displacement Δr of the π-orbital centroid, on the C=C axis, with respect to the nuclear position of the sp^2 atom (identified here as atom l). Before considering the change of charge, we concentrate on topics 2 and 3 with the help of Eqs. (11.10)–(11.12).

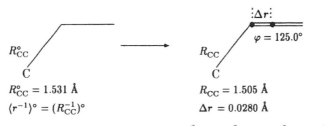

$R_{CC}^\circ = 1.531$ Å

$\langle r^{-1} \rangle^\circ = (R_{CC}^{-1})^\circ$

$R_{CC} = 1.505$ Å

$\varphi = 125.0°$

$\Delta r = 0.0280$ Å

Figure 11.1. The bond transformation $C(sp^3)$—$C(sp^3) \rightarrow C(sp^3)$—$C(sp^2)$.

As regards the latter topics, attention must be given to the indices: F_{kl} mirrors, so to speak, how nucleus k "sees" the electrons of atom l, but F_{lk} represents nucleus l in the field of the electrons of atom k. The sum $F_{kl} + F_{lk}$ gives the total F for that k—l bond.

The relevant bond distance R_{kl}° of the initial molecule and the geometry of the final product, namely, R_{CC} and φ (see Fig. 11.1), are known from the problem. The actual calculation involves four steps: (1) calculating Δr as indicated earlier; (2) applying the theorem (11.11), calculate the distance between the centroid of the π orbital on atom l and the nucleus k; (3) then using the monopole approximation (11.10) to obtain $\langle r_{kl}^{-1} \rangle$, the average inverse distance between the electrons on l and nucleus k; and finally (4) using this $\langle r_{kl}^{-1} \rangle$ result in Eq. (11.12) and finding F_{kl}. $\langle r_{kl}^{-1} \rangle^\circ$ is usually known from the context of the problem [as in the example depicted in Fig. 11.1, where $\langle r_{kl}^{-1} \rangle^\circ = (R_{kl}^{-1})^\circ$] or must be calculated following steps 1–3. Note that step 3 amounts to writing

$$\langle r_{kl}^{-1} \rangle \simeq \frac{1}{4} \left[\frac{3}{R_{kl}} + \frac{1}{\sqrt{R_{kl}^2 + (\Delta r_{\pi l})^2 - 2R_{kl}\, \Delta r_{\pi l} \cos \varphi}} \right] \tag{11.13}$$

Regarding step 1, we approach the problem by observing the following guidelines suggested by direct computations of Δr.

For σ systems (sp^3 carbons and hydrogen), the charges are taken at their respective nuclear positions. This is part of the monopole approximation (11.10). For the aryl carbon atoms, we have $\Delta r = 0$ for obvious symmetry reasons. This leaves us with the olefinic double bonds. They are dealt with in the following manner.

The idea is to find representative $C(sp^2)$—$C(sp^3)$ and $C(sp^2)$—H bonds for general use in applications to olefins, that is, to find an acceptable approximation that bypasses lengthy computations of Δr for each molecule of interest. Let us first evaluate the risks. Consider the $C(sp^2)$—$C(sp^3)$ bond of tetramethylethylene, with $\Delta r = 0.0280$ Å, formed from that of ethane (Fig. 11.1) with $F = -4.30$ kcal/mol (details are given in Example 11.1). Consider also a standard $C(sp^2)$—H bond, calculated for $\Delta r = 0.0290$ Å and $\varphi = 121.7°$, formed from that of ethane with $F = -1.87$ kcal/mol^{-1} (Example 11.2). Now consider propene. On the side of the sp^2 carbon carrying the methyl substituent, direct calculations indicate that $\Delta r = 0.0299$ Å. This $C(sp^2)$—$C(sp^3)$ bond is formed from that of ethane with $F = -4.53$ kcal/mol, which is 0.23 kcal/mol *more* negative than that of tetramethylethylene. At the other end of the double bond of propene, in turn, the displacement is reduced to 0.0281 Å and the total F of the two hydrogens at that end is -3.53 kcal/mol. For the third CH bond, that formed by the carbon bearing the methyl group F is -1.76 kcal/mol. The total F for the three CH bonds is thus -5.29 kcal/mol, which is 0.32 kcal/mol *less* negative than 3 times the assumed reference, $3(-1.87) = -5.61$ kcal/mol. Briefly, the price paid for this approximation is a discrepancy of 0.09 kcal/mol between the approximated and the direct results obtained for propene. The selection of a standard value of 0.0290 Å, for the calculation of $C(sp^2)$—H bonds seem reasonable—that of ethylene itself is 0.0292 Å—as well as the selection of the symmetric tetramethylethylene molecule, with $\Delta r = 0.0280$ Å, for the evaluation of a model $C(sp^2)$—$C(sp^3)$ bond. These rules are applied in all the following calculations of olefins.

Examples. Let us now work out some examples. The energies are given in kcal/mol using the conversion factor 1 hartree = 627.51 kcal/mol. Distances and inverse distances are indicated in Å and Å^{-1}, respectively, with 1 bohr = 0.52917 Å. The angles φ are reported in degrees. The calculations are made for reference charges, namely, 35.1, 7.7, and 13.2 me for sp^3, ethylenic, and aryl carbons, respectively.

Example 11.1. We begin with the CC bond of ethane, $R_{kl}^\circ = 1.531$, assuming $\langle r_{kl}^{-1} \rangle^\circ = (R_{kl}^{-1})^\circ$ for this σ system. This bond is transformed into the $C(sp^3)$—$C(sp^2)$ bond of tetramethylethylene, with $R = 1.505$, $\varphi = 125.0$ (experimental geometry) and $\Delta r = 0.0280$ (from a direct computation of Δr). The cosine theorem (11.11) tells us that the centroid of the π orbital is at a distance of 1.521233 from the nucleus of the sp^3 carbon. Using this result in (11.10), we get $\langle r_{kl}^{-1} \rangle = 0.66268$. Finally, Eq. (11.12) gives $F_{kl} = -4.076$. For the calculation of F_{lk} (i.e., for the electrons of the methyl carbon in the field of the nuclear charge of the sp^2 carbon), we have $\langle r_{kl}^{-1} \rangle^\circ = (R_{kl}^{-1})^\circ$, $\langle r_{kl}^{-1} \rangle = R_{kl}^{-1}$ and $F_{lk} = -0.225$. The total F for this transformation is thus $F = -4.30$.

Example 11.2. We transform a H—$C(sp^3)$ bond, $R_{kl}^\circ = 1.08$, into a H—$C(sp^2)$ bond, $R_{kl} = 1.08$. with $\Delta r = 0.0290$. (The sp^2 carbon is atom l.) For ethylene and, more generally, for CH bonds in *trans*-olefins, $\varphi \simeq 121.7$. Using this angle, we get $\langle r_{kl}^{-1} \rangle = 0.92264$. For the original CH bond it is $\langle r_{kl}^{-1} \rangle^\circ = (R_{kl}^{-1})^\circ$. Thus we obtain $F_{kl} = F = -1.87$ because $F_{lk} = 0$. For *cis*-olefins, we use $\varphi = 117.4$, the angle calculated for *cis*-butene, and get $F = -1.67$.

Example 11.3. To make things a little more interesting, we transform the $C(sp^3)$—$C(sp^2)$ bond of tetramethylethylene (Example 11.1) into a $C(sp^2)$—$C(sp^2)$ single bond like that found in 1,3-butadiene, assuming $R = 1.49$, $\varphi = 123$, and

$\Delta r = 0.028$. The latter three entries give $\langle r_{kl}^{-1}\rangle = 0.66942$. Now we calculate F_{kl}, where l is the sp^2 carbon of Example 11.1; thus we use $R^\circ = 1.505$ and $\langle r_{kl}^{-1}\rangle^\circ = 0.66268$ here. Equation (11.12) gives $F_{kl} = 0.086$. F_{lk} represents the sp^3 carbon of the original bond that is transformed into an sp^2 carbon, again with $\langle r_{kl}^{-1}\rangle = 0.66942$, but with $\langle r_{kl}^{-1}\rangle^\circ = (R_{kl}^{-1})^\circ$. So we obtain $F_{lk} = -3.940$ and $F = F_{kl} + F_{lk} = -3.85$.

Example 11.4. Here we use the $C(sp^3)$—$C(sp^2)$ bond of tetramethylethylene and transform it into a $C(Ar)$—$C(sp^2)$ bond like that found in styrene, $R = 1.445$, $\varphi = 126$ (calculated geometry) with $\Delta r = 0.028$. Let l be the $C(sp^2)$ atom of the original bond, with $R_{kl}^\circ = 1.505$ and $\langle r_{kl}^{-1}\rangle^\circ = 0.66268$ (see Example 11.1). For styrene we get $\langle r_{kl}^{-1}\rangle = 0.69007$ and $F_{kl} = -0.575$. For the aryl carbon atom and its electrons, it is $\langle r_{kl}^{-1}\rangle^\circ = (R_{kl}^{-1})^\circ$ and $\langle r_{kl}^{-1}\rangle = R_{kl}^{-1}$. Remembering that $q_k = 13.2 \times 10^{-3}$ e, we get $F_{lk} = -0.207$ and, finally, $F = -0.78$ for this transformation.

It is clear that some information is lost as a result of our approximation, specifically regarding the individual bonds formed by sp^2 carbon atoms, but the total of the F values in a molecule is expected to be generally reasonably accurate. The salient feature of ethylenic double bonds, namely, the inward displacement of π orbital centroids on the C=C axis revealed by direct calculations, and its important role in energy calculations, can now be put in a clear perspective and efficiently tested for large collections of molecules.

11.2 UNSATURATED SYSTEMS

Unsaturated hydrocarbons, such as alkenes, polyenic material, as well as aromatic molecules, are our target. Carbonyl compounds are considered in Chapter 16.

The a_{kl} parameters are deduced from Eq. (10.41), but attention must be given to the fact that σ and π populations vary in inverse directions, Eq. (6.3), a circumstance that affects a_{kl} and thus $a_{kl}\Delta q_k$.

The CC and CH bonds of ethane (Example 10.1), and the final selection $\varepsilon_{CC}^\circ = 69.633$ and $\varepsilon_{CH}^\circ = 106.806$ kcal/mol, are used to get the CC and CH bonds found in unsaturated hydrocarbons by retaining both the contribution of F_{kl}, Eq. (11.12), and the effect of charge variations described by Eq. (10.37). The reference CC double bond of ethylene and the reference CC bonds of benzene, however, roughly estimated along the lines described in Example 10.1, are deduced from the appropriate CH bond energies and the energy of atomization of the corresponding molecule, ΔE_a^*, obtained from experimental data.

Examples 10.1 and 11.1–11.4 are used, along with a number of similar examples presented below, to generate all the required $C(sp^2)$—$C(sp^3)$, $C(sp^2)$—$C(sp^2)$, and $C(sp^2)$—H bonds, including those formed by aryl carbons. Conjugation must therefore also be considered.

These topics are developed as follows.

Calculation of π Systems

Consider the exact definition of a_{kl} from Eq. (10.32). When atom k is a sp^2 carbon, we can safely neglect the second- and higher-order terms because the Δq_k values are small, in favor of the simple approximation, Eq. (10.41). However, we must consider both σ- and π-electron densities and their variations. The appropriate first derivatives $(\partial E_k^{vs}/\partial N_k)°$ are indicated in Table 10.3.

Moreover, considering the shift of the π-orbital centroid, $\langle r_{kl}^{-1}\rangle°$ also depends on whether we are describing σ or π electrons. Briefly, two a_{kl} occur in unsaturated systems, namely, a_{kl}^{σ} and a_{kl}^{π} for the σ and π electrons, respectively.

Separating the variations of the σ charges Δq_k^{σ} from those of the π charges Δq_k^{π}, we write

$$a_{kl}\Delta q_k = a_{kl}^{\sigma}\Delta q_k^{\sigma} + a_{kl}^{\pi}\Delta q_k^{\pi} \tag{11.14}$$

On the other hand, we have $\Delta q_k = \Delta q_k^{\sigma} + \Delta q_k^{\pi}$, meaning that the a_{kl} of Eq. (11.14) is simply the weighted average of a_{kl}^{σ} and a_{kl}^{π}:

$$a_{kl} = \frac{a_{kl}^{\sigma}\Delta q_k^{\sigma} + a_{kl}^{\pi}\Delta q_k^{\pi}}{\Delta q_k^{\sigma} + \Delta q_k^{\pi}} \tag{11.15}$$

It turns out that the variations of σ and π charges are not independent variables. They obey, at least to a good approximation, a simple relationship [Eq. (6.3)]:

$$\Delta q_k^{\sigma} = m\Delta q_k^{\pi}$$

where m (<0) describes how σ populations at atom k increase when π populations decrease, and vice versa. Combining the latter equation with (11.15), we find

$$a_{kl} = \frac{m\,a_{kl}^{\sigma} + a_{kl}^{\pi}}{m + 1} \tag{11.16}$$

The technical difficulty encountered in applications of Eq. (11.16) comes from the fact that calculated σ charges appear to be considerably more basis-set-dependent than are π charges, rendering the evaluation of m uncertain. Presently, the best estimates are $m \simeq -0.955$ for ethylenic molecules and $m \simeq -0.814$ for benzenoid hydrocarbons [129], but these estimates should be taken with a grain of salt. It would be welcome if someone could figure out how to get better results.

Conjugation

The modifications taking place in CC and CH bonds due to the replacement of an sp^3-hybridized carbon by an sp^2 carbon atom and the explicit introduction of the $\sigma-\pi$ separation are important in applications of the energy formula (10.33) to cover olefins.

At this point we have considered the merits of the function F and dealt with monoolefins applying the same scheme as used for paraffins. In applications to polyunsaturated systems, such as 1,3-butadiene and aromatic hydrocarbons, it appears that the function F does not translate all possible geometry-related effects. Attention must be given to an important property of π systems—conjugation—which is introduced as follows, taking butadiene as an example.

The electron diffraction investigation of the molecular structure of 1,3-butadiene indicates that the planar *trans* form is predominating; in fact, no sign of other conformations was observed [223], a result that is convincingly supported by ab initio MO theory [224], at the level of minimal (STO-3G) and extended [4-31G, 6-31G, 6-31G*, $(7s\,3p|3s)$ and larger] basis set calculations. Structural parameters were determined [223,225] namely, $R(C—C) = 1.463 \pm 0.003$, $R(C=C) = 1.341 \pm 0.002$, and $R(C—H) = 1.090 \pm 0.004$ Å. While the double bond is, in essence, that of ethylene itself (a fact that prompts us to treat all double bonds on the same footing), the single CC bond of 1,3-butadiene is significantly shorter than the "usual" single bonds, that is, those formed by two sp^3 carbon atoms. The short single bond and the planar equilibrium conformation have been attributed to π-electron delocalization, which gives the single bond some amount of double-bond character and is most effective in the planar molecule [226]. It became apparent, however, that delocalization was less than predicted by the simple Hückel method [227]; the CC bond shortening was then attributed to the change in hybridization of the bond-forming atoms [228]. These points merit consideration in the interpretation of anticipated energy effects resulting from steric constraints that would force noncoplanarity of the double-bond system.

Extended basis set MO calculations indicate that, indeed, resonance is the main factor responsible for the shortening of the CC single bond in the planar *s-trans* conformation of 1,3-butadiene [229,230]. In this conformation the distinction between σ and π electrons is evident. For the perpendicular case, on the other hand, it is always possible to define one localized π MO on each C=C fragment. The SCF localized π MO located on one of the C=C moieties has σ tails on the other C=C fragment and the σ MOs of each C=C moiety have components on the $2p_y$ atomic orbitals associated with the other fragment. This effect, known as *hyperconjugation*, has been found to stabilize the perpendicular form, largely compensating for the lack of π conjugation, and leads to a similar shortening of the central CC bond [230].

Ab initio analyses thus substantiate the concept of π conjugation in the planar conformation and extend it to hyperconjugation in perpendicular forms. Direct calculations of resonance energies lead to estimates in the neighborhood of $\sim$10 kcal/mol [224,230]. Resonance energy is now defined in its original sense as the energy difference between the conjugated system and its reference state without resonance; the latter is represented by a model wavefunction consisting of strictly localized nonresonating π molecular orbitals. Hyperconjugation energy is defined in a similar manner [230]. The hyperconjugative stabilization in the perpendicular form at its SCF optimum distance ($\sim$1.5 Å) turns out to be almost as large ($\sim$8.0 kcal/mol) as the π conjugation energy in the equilibrium planar conformation ($\sim$10.4 kcal/mol at $\sim$1.48 Å) [230]. This change from resonance to hyperconjugative stabilization of the central CC bond, however, is not reflected in the function F, nor is it in the $\sum_k\sum_l a_{kl}\Delta q_k$ term because the latter considers only the effects of changing electron populations at the bond-forming atoms under otherwise identical conditions. Rather, the change in geometry from a nonplanar to a planar form that is at the origin of the gain in conjugation at the expense of a hyperconjugative stabilization should be accounted for and accommodated by an appropriate change of the reference bond energy ε_{CC}° of the central CC single bond.

The extent of the change in ε_{CC}°, however, is difficult to evaluate theoretically. A valuable rough estimate is offered by the SCF results discussed above [230] indicating that hyperconjugative stabilization is less effective (by $\sim$2.4 kcal/mol) than π conjugation. This estimate is similar to the familiar thermochemical resonance energy of some 3 kcal/mol [231] and suggests that the reference bond energy ε_{CC}° in a planar resonating arrangement should be larger by approximately that amount relative to that applicable in a markedly nonplanar situation. In practical applications, a uniform value of $\sim$2.8 kcal/mol seems a reasonable, but perforce approximate, compromise.

While π conjugation and hyperconjugation are certainly important contributions to the thermochemical stability of CC single bonds embedded in double-bond systems, molecules featuring this arrangement still appear to be strongly localized (i.e., poorly conjugated). Of course, this fact is well known from simple Hückel theory with equal CC resonance integrals leading to alternating π-bond orders. In this sense, keeping in mind the features linked to π conjugation versus hyperconjugation revealed by SCF theory, it remains valid to apply our energy formula (10.33) to conjugated systems with the understanding that resonance effects must be taken into consideration.

Reference Bonds

Here we start off with the CC and CH reference bond energies of ethane and use them to get the CC and CH bonds occurring in unsaturated hydrocarbons by considering both the contribution of F [Eq. (11.12)] and the change of charge given by $a_{lk}\Delta q_l$ [Eq. (10.37)]. The new reference bonds thus obtained are indicated in Table 11.2.

The appropriate transformations are described by Examples 11.5–11.11. The units are those used for Examples 10.1 and 11.1–11.4. The a_{lk} parameters, in kcal mol^{-1}

TABLE 11.2. Selected Reference Bonds and Their Energies (kcal/mol)

Bond	Type	Occurrence	R_{kl} (Å)	q_k° (me)	q_l° (me)	Energy
ε_1°	$C(sp^3)$—$C(sp^3)$	Ethane	1.531	35.1	35.1	69.63
ε_2°	$C(sp^2)$—$C(sp^2)$	Ethene	1.34	7.7	7.7	139.37
ε_3°	$C(Ar)\!:::\!C(Ar)$	Benzene	1.397	13.2	13.2	115.39
ε_4°	$C(sp^3)$—$C(sp^2)$	Olefins	1.505	35.1	7.7	77.67
ε_5°	$C(sp^2)$—$C(sp^2)$	Butadiene	1.49	7.7	7.7	89.14
ε_6°	$C(Ar)$—$C(sp^3)$	Toluene	1.48	13.2	35.1	79.33
ε_7°	$C(Ar)$—$C(sp^2)$	Styrene	1.445	13.2	7.7	89.69
ε_8°	$C(Ar)$—$C(Ar)$	Aromatics, conjugated	1.397	13.2	13.2	91.21
ε_9°	$C(Ar)$—$C(Ar)$	Biphenyl	1.49	13.2	13.2	88.89
ε_{10}°	$C(sp^3)$—H	Ethane	1.08	35.1	-11.7	106.81
ε_{11}°	$C(sp^2)$—H	Ethene	1.08	7.7	-11.7	110.69
ε_{12}°	$C(Ar)$—H	Benzene	1.08	13.2	-11.7	111.41

me^{-1} units, are for the interatomic distance specified in parentheses, in Å. This distance is that of the *initial* bond because the change in geometry is part of F. The derivative $(\partial E_l / \partial N_l)° = -0.375$ au, on the other hand, is that of the newly formed sp^2 carbon ($=$carbon l). This approach implicitly maximizes the association of all energy changes resulting from the rehybridization of a carbon atom with the multiple bond created during this rehybridization. In this scheme, the gain in electronic charge (e.g., -27.4 me) in going from the ethane to the ethylene carbon atom, is considered to take place at the σ level; the fact that a π orbital "separates out" with a concurrent change of the atom's "own" energy is regarded as part of the newly formed multiple bond and its energy. The a_{lk} parameters are easily obtained from Eq. (10.41) remembering that $v = 3$ for the sp^2 carbon atoms.

Example 11.5. $\varepsilon_4°$ is obtained from $\varepsilon_1°$ as $\varepsilon_4° = \varepsilon_1° + a_{C'C}°(7.7 - 35.1) + F$, with $a_{C'C}°(1.531) = -0.450_3$ and $F = -4.30$, calculated in Example 11.1.

Example 11.6. $\varepsilon_5°$ follows from $\varepsilon_4°$, i.e. $\varepsilon_5° = \varepsilon_4° + a_{C'C}°(7.7 - 35.1) + F +$ conjugation, with $a_{C'C}°(1.505) = -0.456_7$ and $F = -3.85$ (see Example 11.3). Conjugation is accounted for as indicated earlier and is tentatively estimated at 2.8 kcal/mol.

Example 11.7. The C(Ar)—C(sp^3) bond like that found in toluene is given by $\varepsilon_6° = \varepsilon_1° + a_{C'C}°(13.2 - 35.1) + F$, with $a_{C'C}°(1.531) = -0.450_3$. F is calculated by observing that $\langle r_{kl}^{-1} \rangle° = (R_{kl}^{-1})°$ for ethane and $\langle r_{kl}^{-1} \rangle = R_{kl}^{-1}$ for toluene, with $r_{C'C} = 1.48$ (calculated geometry), which gives $F = -0.169$ and $\varepsilon_6° = 79.33$.

Example 11.8. The conjugated CC single bond like that of styrene is deduced from $\varepsilon_4°$ as $\varepsilon_7° = \varepsilon_4° + a_{C'C}°(13.2 - 35.1) + F +$ conjugation, where $a_{C'C}°(1.505) = -0.456_7$. $F = -0.78$ (Example 11.4), and conjugation adds 2.8 kcal/mol.

Example 11.9. $\varepsilon_8°$ denotes a conjugated carbon–carbon single bond formed by two aryl sp^2 carbon atoms, at a distance 1.397 (which is the experimental distance in benzene [232]). This bond is calculated from $\varepsilon_1°$, that is, $\varepsilon_8° = \varepsilon_1° + 2a_{C'C}° \times (13.2 - 35.1) + F +$ conjugation, with $a_{C'C}°(1.531) = -0.450_3$. Finally, $F = -0.94_2$ is obtained by observing that $\langle r_{kl}^{-1} \rangle° = (R_{kl}^{-1})°$ and $\langle r_{kl}^{-1} \rangle = R_{kl}^{-1}$, with $R_{kl}° = 1.531$ and $R_{kl} = 1.397$.

Example 11.10. $\varepsilon_9°$ represents a nonconjugated CC bond between aryl carbon atoms, like that of biphenyl, with $R = 1.49$ [233], $F = -0.14$. Remembering that $\varepsilon_6°$ was obtained from $\varepsilon_1°$ by replacing one CH$_3$ of ethane by a phenyl group, the central CC bond of biphenyl is approximated as $2(\varepsilon_6° - \varepsilon_1°) + \varepsilon_1° + F = 88.89$.

Example 11.11. The reference energies of C(sp^2)—H bonds are deduced from $\varepsilon_{10}°$ namely, $\varepsilon_{11}° = \varepsilon_{10}° + a_{C'H}°(7.7 - 35.1) + F$ with $a_{C'H}°(1.08) = -0.210_2$ and $F = -1.87$ (Example 11.2), whereas $F = 0$ for $\varepsilon_{12}° = \varepsilon_{10}° + a_{C'H}°(13.2 - 35.1) = 111.41$.

The reference bond energies reported in Table 11.2 faithfully mirror the constraints specified by Eq. (10.34) and described in Examples 10.1 and 11.1–11.11. Although covering a wide range of possible situations, their interrelations turn out to be very simple in the end; clearly established, they are shown to advantage through straightforward uncomplicated applications of Eqs. (10.36), (10.37), and (11.12). They could perhaps be improved for general use. But we take them as indicated, for a good reason—as they are, they lead to atomization energies and standard enthalpies of formation that are in excellent agreement with their experimental counterparts, usually within experimental uncertainties.

The a_{kl} Parameters

Let us now develop the a_{kl} parameters of Eq. (10.37).

With sp^2 carbons, whose charges vary concurrently at the σ and π levels, $\Delta q_k^\sigma = m \Delta q_k^\pi$, the $a_{kl} \Delta q_k$ term becomes $a_{kl}^\sigma \Delta q_k^\sigma + a_{kl}^\pi \Delta q_k^\pi$; that is, according to Eq. (11.16), we obtain

$$a_{kl} = \frac{m a_{kl}^\sigma + a_{kl}^\pi}{m + 1}$$

Present knowledge about atomic charges of sp^2 systems is unfortunately insufficient; it does not permit direct evaluations of m. Extensive numerical analyses, namely, comparisons of experimental atomization energies with their all-theoretical formulation leaving m as the only unknown parameter, were thus carried out. They indicated that [109,129,192]

$$
\begin{aligned}
m &\simeq -0.955 &&\text{for olefins} \\
&\simeq -0.814 &&\text{for nonsubstituted aromatic C atoms} \\
&\simeq -0.90 &&\text{for substituted aromatic C atoms}
\end{aligned}
$$

(By "nonsubstituted aromatic carbons," we mean C atoms not engaged in a bond with a nonaromatic carbon, as in toluene, for example.)

Now we turn to Eq. (10.41) for a_{kl}. The derivatives $(\partial E_k / \partial N_k)^\circ$ are listed in Table 10.3.

The average inverse distance between the electrons on k and nucleus Z_l (transcribed here by R_{kl}, in Å) were taken at $R_{CH} = 1.08$ (σ electrons of sp^3 or sp^2 C), $R_{CH} = 1.10$ (π electrons of sp^2 carbon), $R_{CC} = 1.53$ (sp^3–sp^3 bonds), $R_{CC} = 1.34$ (sp^2 σ electrons), and 1.30 Å (sp^2 π electrons) for carbon–carbon double bonds; $R_{CC} = 1.46$ (sp^2–sp^2 single bond) for σ electrons, and, 1.48 Å for the π electrons of that bond. Finally, $R_{CC}(sp^2$–$sp^3) = 1.53$ for σ electrons and 1.552 for the π electrons of the sp^2 carbon atom. For the aromatic carbons, $R_{CC} = 1.40$ for the distance between aryl carbons and $R_{CH} = 1.08$ for C(Ar)—H.

Example 11.12. Typical values are $a_{CC}^\sigma(1.40) = -0.485_0$ and $a_{CC}^\pi(1.40) = -0.461_4$ kcal mol^{-1} me^{-1} for the aromatic CC bonds, $a_{CH}^\sigma(1.08) = -0.210_2$, and $a_{CH}^\pi(1.08) = -0.186_6$ kcal mol^{-1} me^{-1} for the CH bonds. Using Eq. (11.16) with

TABLE 11.3. Selected a_{kl} Parameters (kcal mol^{-1} me^{-1})

Bond kl	m	$a_{kl}^{\sigma\pi}$
$C(sp^3)$—$C(sp^3)$	—	-0.488
$C(sp^3)$—$C(sp^2)$	—	-0.488
$C(sp^2)$—$C(sp^3)$	-0.955	0.275
$C(sp^2)$—$C(sp^2)$	-0.955	0.258
$C(sp^2)$=$C(sp^2)$	-0.955	-0.183
$C(Ar)$—$C(Ar)$	-0.814	-0.358
$C(Ar)$—$C(sp^3)$	-0.90	-0.226
$C(sp^3)$—H	—	-0.247
$C(sp^2)$—H	-0.955	0.454
$C(Ar)$—H	-0.814	-0.083
H—$C(sp^3, sp^2)$	—	-0.632

$m = -0.8137$, one finds $a_{CC}^{\sigma\pi} = -0.358_3$ and $a_{CH}^{\sigma\pi} = -0.083_5$ kcal mol^{-1} me^{-1} for these bonds. On the other hand, for the $C(Ar)$—$C(sp^3)$ bond like that occurring in toluene, we have for the aryl carbon $a_{CC}^{\sigma}(1.48) = -0.463_1$, $a_{CC}^{\pi}(1.48) = -0.439_4$ and hence $a_{CC}^{\sigma\pi} = -0.226$ kcal mol^{-1} me^{-1} for $m = -0.9$.

Although admittedly rough, these approximations turn out to give reasonably accurate final results in comparisons of energy calculations with experimental data. The final a_{kl} data are listed in Table 11.3.

Calculation of $m \simeq -0.90$. For an aromatic carbon linked to a nonaromatic C, $\sum_l a_{kl}$ is $2a_{CC}^{\sigma\pi}(Ar) + a_{CC}^{\sigma\pi}(nAr)$, where $a_{CC}^{\sigma\pi}(Ar)$ is used for benzenic bonds and $a_{CC}^{\sigma\pi}(nAr)$ is for the bond formed with the nonaromatic carbon. The change in charge at the aromatic C is Δq_C (relative to benzene), contributing $\Delta q_C \sum_l a_{kl}$ to ΔE_a^*. On the other hand, this Δq_C is also part of $\sum q_C$, hence of $\sum \Delta q_H$, and contributes $-\Delta q_C \times a_{HC}$. The total contribution of Δq_C is thus

$$[2a_{CC}^{\sigma\pi}(Ar) + a_{CC}^{\sigma\pi}(nAr) - a_{HC}]\Delta q_C$$

Of course, in this calculation, Δq_C must *not* be included in the evaluation of Δq_H. Using Eq. (11.16), as well as $\Delta q_C = (m+1)\Delta q_C^\pi$ and $\delta = 157 \, \Delta q_C^\pi$, we find that the energy contributed by Δq_C is (in au),

$$m[2a_{CC}^{\sigma}(Ar) + a_{CC}^{\sigma}(nAr) - a_{HC}]\frac{\delta}{157} + [2a_{CC}^{\pi}(Ar) + a_{CC}^{\pi}(nAr) - a_{HC}]\frac{\delta}{157}$$

where δ is the ^{13}C shift relative to benzene. Inserting here the appropriate a_{kl}'s (in au) one obtains the energy contributed by Δq_C, namely (in kcal/mol)

$$-(1.2766\,m + 1.1636)\left(\frac{627.51}{157}\right)\delta = -0.06_0\,\delta \qquad (11.17)$$

for $m = -0.8997$. It is clear that the final result depends heavily on m, far beyond the precision of present SCF charge analyses. Inspection of molecules corresponding to this situation has consistently led to this -0.06_0 parameter; hence our choice $m = -0.90$. The selection of $R_{CC} = 1.397$ for $a_{CC}(Ar)$ and $R_{CC} = 1.48$ for $a_{CC}(nAr)$ offers the best results overall.

Additional features need be considered in actual numerical calculations of benzenoid hydrocarbons, such as questions pertaining to the planarity (or lack of it) of complex structures. They are presented in Chapter 14.

11.3   RECAPITULATION

The Hellmann–Feynman theorem offers a convenient way to highlight the main features of chemical binding. By choosing the nuclear charges as parameters, it becomes possible to define the binding of each individual atom in a molecule without having recourse to an explicit partitioning of that molecule into atomic subspaces.

Thus we know about "atoms in a molecule" uniquely defined by the potentials at their nuclei. By the same token, explicit calculations of the electronic kinetic energies and of all two-electron integrals are avoided.

Consideration of the valence region of a ground-state molecule leads to a formula for the energy of atomization in an approximation akin to the Thomas-Fermi model. The final result, from Eq. (10.33), is

$$\Delta E_a^* = \sum_{k<l} \varepsilon_{kl}^\circ + \sum_k \sum_l a_{kl}\Delta q_k + F - E_{nb}$$

No set of invariant, empirically determined bond energies $\varepsilon_{kl}^{empir.}$ can ever describe electroneutral molecules. But electroneutrality must be preserved. This is achieved by the $\sum_k\sum_l a_{kl}\Delta q_k$ term. In passing, we should keep in mind that these bond energies are extremely sensitive to the atomic electronic charges; as revealed by the a_{kl} parameters of Table 10.4, a gain of ~ 1 me at the ends of a sigma bond augments its energy by ~ 1 kcal/mol, which is a lot to reckon with. On the other hand, only most precise charge results are of any use in bond energy calculations. One may thus rightfully wonder whether the a_{kl} parameters propounded here are sufficiently accurate. A few words about this matter are in order.

The $a_{kl}\Delta q_k$ term describes how the nuclear–electronic, interelectronic, and electronic kinetic energies change along with the charge on atom k bonded to atom l. Its $(1/v_k)[(\partial E_k/\partial N_k)^\circ \Delta q_k + \cdots]$ part pertains solely to atom k. The accompanying change of nuclear–electronic attraction due to the presence of nucleus Z_l is $Z_l^{eff} \times \Delta q_k/R_{kl}$, which, after multiplication by $\frac{3}{7}$, accounts for the corresponding changes of interelectronic and kinetic energies as well. Any departure from constancy of the Thomas–Fermi ratio $\frac{3}{7}$, which is unlikely to exceed $\pm 3\%$, affects just that part of $a_{kl}\Delta q_k$. On the other hand, $\sum_k\sum_l a_{kl}\Delta q_k$ represents only a small part, approximately 3–4% of ΔE_a^*. This explains that, all things considered, our present use of the Thomas–Fermi approximation is no cause of concern in terms of the accuracy of the method.

The reason for F is not too far to seek. In the general context of charge distributions, the role of an atom cannot be described solely by reference to the number of electrons it contributes to the molecular fabric. The centroid of its integrated charge density is equally important.

This is where F [Eq. (11.12)] comes in. The approximations that were introduced partly to gain physical insight and partly to facilitate numerical applications rest on a simplifying assumption, namely, the point-charge model for the saturated hydrocarbons or, more precisely, on the approximate validity of Eq. (11.5). These approximations cannot be proved a priori to be of minor import. However, the contamination introduced with their use is certainly minimal because they affect only a minor part ($<5\%$) of the dissociation energies to be calculated. The final results, namely, comparisons between calculated and experimental energies, provide an a posteriori argument for claiming that these approximations are in fact very good ones. But then they must be interpreted as general rules that the partitioning of electron densities obtained from accurate electronic wavefunctions must obey. This is especially important in any extension of an energy theory to cover olefins as well as paraffins; for that extension requires explicit introduction of the $\sigma-\pi$ separation. As a consequence, the point-charge approximation must be reformulated as a "principle" valid for the averages of electrostatic potentials of the σ orbitals of any given atom in a hydrocarbon molecule. This has two distinct practical advantages. It suggests a simple numerical correction that allows applications to ethylenic hydrocarbons of the same scheme (in terms of local charges) as used for paraffins and assigns that correction an extremely simple physical interpretation (that of a π centroid displacement) amenable to numerical evaluation from accurate ab initio computations. On the other hand, it involves three important conceptual features: (1) the notion of in situ atomic orbitals, which must now be introduced explicitly, (2) the postulate that there is a Mulliken-type partitioning of the molecular electron density into atomic orbital contributions that obey the "principle of orbital balancing," and (3) the implication that a CH bond or a CC bond in an olefin posseses the same basic properties as in paraffins, except for the effective net charges of the atoms involved. The remarkably effective orbital balancing described here signals a general propensity of the local electronic structures to resist modifications and suggests the idea that atomic charge variations should perhaps also be regarded as events occurring most reluctantly.

Coming now to conjugation, results like those due to Malrieu et al. play an important role in the discussion of polyunsaturated systems. They offer the theoretical background for the well-known thermochemical stabilization of a chemical bond due to conjugation. This stabilization is not reflected in the $\sum_k \sum_l a_{kl} \Delta q_k + F$ part of Eq. (10.33). It must be included in the definition of the relevant reference bond energy. This formulation, which is the simplest possible one, is adequate because it turns out that, for all practical purposes, the amount of conjugation associated with a given bond can be treated as a constant (representing a rough but acceptable approximation) unless, of course, changes of molecular structure force a disruption of conjugation. This causes no difficulties in applications of (10.33), but requires some advance knowledge about molecular structure.

The flexibility and internal consistency of the present theory are well illustrated by the transformations that generate the sets of parameters required for the unsaturated hydrocarbons from those of their saturated models. But most importantly they preserve the original form and great simplicity of the basic bond energy formula, $\varepsilon_{kl} = \varepsilon_{kl}^{\circ} + a_{kl}\Delta q_k + a_{lk}\Delta q_l$, as well as its accuracy.

This bond energy theory is to some extent a further step toward solving a persistent problem of theoretical chemistry; bridging the gap between the apparent simplicity of observed molecular behavior and the intricacies and ambiguities that plague the translation of accurate quantum-mechanical results into simple chemical concepts and rules. The remarkable accuracy with which our formulas allow prediction of atomization energies is a strong indication that hidden regularities can be discovered as a result of patient processing of general theory, but the price to be paid is that the use of final equations like our bond energy formula often rests on simplifying assumptions that, as mentioned before, cannot be proved a priori to be of minor import. The agreement within experimental accuracy of many predicted and experimental results provides an a posteriori argument for claiming that these approximations are satisfactory. The internally coherent picture thus obtained provides a strong indication that the analysis of accurate ab initio computations into models, rules, properties, and interpretations meaningful to chemists should be carried out along the lines emerging from this work.

CHAPTER 12

BOND DISSOCIATION ENERGIES

12.1 SCOPE

Consider a polyatomic molecule and focus attention on a particular atom pair, k and l, forming a bond with intrinsic energy ε_{kl}. This energy cannot be observed in isolation. What can be measured (in principle) is D_{kl}, the bond dissociation energy, that is, the energy required to break up that bond. Now, D_{kl} depends on a number of events accompanying bond breaking, including possible geometry and hybridization changes affecting the fragments. Briefly, $D_{kl} \neq \varepsilon_{kl}$ in polyatomic molecules.[1] It is understood that D_{kl}, like ε_{kl}, refers to molecules at their potential minimum.

Intrinsic bond energies and bond dissociation energies meet different practical needs. The former play an important role in the description of ground-state molecules. Dissociation energies come into play when molecules undergo reactions. Now, any interaction between a molecule and its environment (such as complex formation or adsorption onto a metallic surface, for example, or hydrogen bonding) affects its electron distribution and thus the energies of its chemical bonds. If we figure out the relationship between dissociation and intrinsic bond energies, we could begin to understand how the environment of a molecule can promote or retard the

---

[1]For diatomic molecules, of course, $\Delta E_a^* = \varepsilon_{kl} = D_{kl}$. In polyatomic molecules, consideration of a stepwise cleavage of all the bonds gives $\Delta E_a^* = \sum_{k<l} D_{kl}$, which does not mean that the individual dissociation energies are the same as the intrinsic bond energies of the original molecule. In methane, for example, $\varepsilon_{CH} = 104.86$ kcal/mol, but the dissociation energy accompanying the cleavage $CH_4 \rightarrow CH_3{}^\bullet + H$ is $D_{CH} = 111.4$ kcal/mol.

Atomic Charges, Bond Properties, and Molecular Energies, by Sándor Fliszár
Copyright © 2009 John Wiley & Sons, Inc.

dissociation of one or another bond of particular interest in that molecule. This outlook hints at a rich potential of future research exploiting charge analyses to gain insight into bond energies, first, and, going from there, into matters of great import regarding the dissociation of chemical bonds.

The energy required to break up a molecule at equilibrium, say, K–L, into fragments K· and L·

$$K - L \longrightarrow K \cdot + L \cdot$$

with atoms $k \in K$ and $l \in L$, is defined by

$$D_{kl} = \Delta E_a^*(KL) - \Delta E_a^*(K\cdot) - \Delta E_a^*(L\cdot) \tag{12.1}$$

where $\Delta E_a^*(KL)$, $\Delta E_a^*(K\cdot)$, and $\Delta E_a^*(L\cdot)$ are the appropriate ground-state atomization energies of the reactant K–L and of the fragments K· and L·, respectively, in their hypothetical vibrationless state at $0\,K$. (One or both reaction products could be atoms, depending on what K–L is.)

The atoms $k \in K\cdot$ and $l \in L\cdot$ form the bond with intrinsic energy ε_{kl} that keeps the fragments K and L united in the original molecule. We want to determine how D_{kl} and ε_{kl} are related to one another.

12.2   THEORY

Consider the bond-by-bond partitioning of $\Delta E_a^*(KL)$ shown in Eq. (10.2), but rename the bonds as follows

$$\Delta E_a^* = \sum_{t<v} \varepsilon_{tv} - E_{nb} \tag{12.2}$$

to avoid confusion with the k—l bond under scrutiny. When we calculate $\Delta E_a^*(KL)$, the summation in (12.2) obviously collects all existing bonds, including the k—l bond. Here we focus on the k—l bond. So we subdivide the right-hand side of Eq. (12.2) and consider only the bonds found in the molecular subunit K and the non-bonded interactions involving only the atoms belonging to K. The sum of all these contributions is $\Delta E_a^*(K)$. This is the part of $\Delta E_a^*(KL)$ that is associated with the group of atoms of K as it exists in that particular molecule KL prior to its dissociation. We proceed similarly with the subunit L and obtain $\Delta E_a^*(L)$. The sum $\Delta E_a^*(K) + \Delta E_a^*(L)$ thus collects the intrinsic energies of all the bonds found in the original molecule except one, ε_{kl}, and all the nonbonded interactions, except those between the atoms of the subunit K and those of the subunit L, represented by $E_{nb}(K \cdot\cdot L)$. The energy balance satisfying exactly Eq. (12.2) for the original molecule KL is therefore

$$\Delta E_a^*(KL) = \Delta E_a^*(K) + \Delta E_a^*(L) + \varepsilon_{kl} - E_{nb}(K \cdot\cdot L) \tag{12.3}$$

When K and L are identical in the molecule (as in CH_3—CH_3, for example, but not in $HOH \cdots OH_2$), each of these groups is necessarily electroneutral. Electroneutral subunits in a molecule, taken exactly as they are in molecules such as KK and LL, are identified by the superscript "∘" Under these conditions, it follows from Eq. (12.3) that

$$\Delta E_a^*(K^\circ) = \frac{1}{2}[\Delta E_a^*(KK) - (\varepsilon_{kl} - E_{nb}(K \cdot \cdot K))] \tag{12.4}$$

which is useful in numerical applications.

The energy formulas [Eqs. (12.1)–(12.3)] translate straightforward applications of first-principle energy conservation. No hypothesis is made, other than that regarding the assumed validity of separating the molecular binding energy into bonded and nonbonded contributions [Eq. (12.2)]. Equation (12.1) features the dissociation energy D_{kl} and Eq. (12.3), the corresponding intrinsic bond energy ε_{kl}. At last we can examine how these bond terms are related to one another.

The key argument is rooted in a simple observation concerning any dissociation KL → K· + L·. *The individual subunits K and L are in general not electroneutral while they are part of the original host molecule, whereas the corresponding radicals certainly satisfy electroneutrality.* A charge neutralization accompanies the transformations K → K· and L → L·. This constraint solves our problem. A few examples[2] (Table 12.1) help us understand why this argument is important.

TABLE 12.1. Net Charges (me) and ΔE_a^* (K) Energies (kcal/mol) of Methyl, Ethyl, and Isopropyl Groups in Selected Host Molecules

K	Host Molecule	Charge	$\Delta E_a^*(K)$
CH_3	CH_4	9.05	314.53
	CH_3COCH_3	3.60	318.45
	CH_3CH_3	0.00	320.34
	$CH_3C_2H_5$	−2.65	322.15
	$CH_3i\text{-}C_3H_7$	−5.02	323.80
	CH_3OCH_3	−10.47	327.50
C_2H_5	$C_2H_5COC_2H_5$	33.60	591.50
	$C_2H_5CH_3$	2.65	610.81
	$C_2H_5C_2H_5$	0.00	612.78
	$C_2H_5OC_2H_5$	−2.59	615.09
$i\text{-}C_3H_7$	$i\text{-}C_3H_7COi,C_3H_7$	50.30	875.00
	$i\text{-}C_3H_7CH_3$	5.02	903.63
	$i\text{-}C_3H_7Oi\text{-}C_3H_7$	3.55	904.50
	$i\text{-}C_3H_7i\text{-}C_3H_7$	0.00	906.67

[2]The $\Delta E_a^*(K)$ energies were derived from Eq. (12.2) by means of bond-by-bond calculations [Eq. (10.37)] and Del Re's approximation [Eq. (10.3)] for the nonbonded contributions [234,235]. Atomization energies of the host molecules, calculated in precisely the same manner, agree within ∼0.2 kcal/mol (average deviation) with their experimental counterparts.

These examples indicate (1) the net charge of K (i.e., to what extent K departs from exact electroneutrality as long as it is part of its host molecule) and (2) the corresponding response in $\Delta E_a^*(K)$ energy (i.e., how K "feels" this departure from electroneutrality).

This response is surely significant, even for small departures from electroneutrality. A fragment K increases its thermochemical stability when it gains electronic charge from its environment: $\Delta E_a^*(K)$ becomes larger.

Let us now profit by what we have learned from these results and develop our strategy with the help of the examples worked out for the methyl groups. These groups differ from one another in $\Delta E_a^*(K)$ energy, and we know by how much. Therefore, it suffices to know how any one of them differs from the true $CH_3\cdot$ radical in order to gain the same information for the other CH_3 groups. A convenient reference is the neutral CH_3 of ethane because it is isoelectronic with $CH_3\cdot$. Its $\Delta E_a^*(K)$ value is deduced from (12.4) and is written $\Delta E_a^*(K^\circ)$. This choice is convenient because it associates energy changes $\Delta E_a^*(K) - \Delta E_a^*(K^\circ)$ with charge neutralization—hence the term *charge neutralization energy* (CNE).

The idea embodied in the concept of charge neutralization is simple: molecular subunits that are not individually electroneutral in the host molecule must imperatively restore their correct numbers of electrons when dissociation takes place.

Electrons come in whole numbers.

The methyl group of CH_4, for example, is electron-deficient by 9.05 me. When recovering this charge during the cleavage $CH_4 \rightarrow CH_3\cdot + H$, its energy *decreases* by CNE $= -5.81$ kcal/mol, meaning that $\Delta E_a^*(CH_3)$ *increases* by that amount. Similarly, during the dissociation $C_2H_5—CH_3 \rightarrow C_2H_5\cdot + CH_3\cdot$, when the CH_3 of propane restores its electroneutrality by losing its excess electronic charge, -2.65 me, its energy increases by CNE $= 1.81$ kcal/mol. At the same time, the energy of the C_2H_5 moiety that recovers that electronic charge decreases by 1.97 kcal/mol. The total CNE accompanying this bond cleavage is thus -0.16 kcal/mol.

The generalization of these arguments is straightforward. CNE is a pivotal concept that permits us to relate any K in a molecule, described by $\Delta E_a^*(K)$, to the corresponding electroneutral K°, described by $\Delta E_a^*(K^\circ)$. In order to learn how any K embedded in its host molecule differs in energy from the ground-state radical $K\cdot$, it suffices to know once and for all how K° differs from $K\cdot$.

This is the so-called reorganizational energy

$$RE = \Delta E_a^*(K^\circ) - \Delta E_a^*(K\cdot) \tag{12.5}$$

which reflects all possible geometry and hybridization changes accompanied, of course, by significant redistributions of electronic charge. The selection of $\Delta E_a^*(K^\circ)$ as a reference is only one of the possible choices. It is arbitrary but convenient because it makes good sense to first get the number of electrons right, then relax everything else. Having now secured this idea, there is an easy way of exploiting it.

Two radicals are formed in any bond dissociation $KL \rightarrow K\cdot + L\cdot$. For the problem at hand, it is more convenient to consider them jointly rather than proceeding with separate calculations of the individual CNE contributions, which are $\Delta E_a^*(K)-\Delta E_a^*(K^\circ)$ on one hand and $\Delta E_a^*(L) - \Delta E_a^*(L^\circ)$ on the other hand. The sum $\Delta E_a^*(K) + \Delta E_a^*(L)$ differs in principle from that of the corresponding electroneutral fragments, $\Delta E_a^*(K^\circ) + \Delta E_a^*(L^\circ)$. The difference between these sums is CNE:

$$\text{CNE} = \left[\Delta E_a^*(K) - \Delta E_a^*(K^\circ)\right] + \left[\Delta E_a^*(L) - \Delta E_a^*(L^\circ)\right] \tag{12.6}$$

Now, using Eq. (12.3), it follows that

$$\text{CNE} = \Delta E_a^*(KL) - \varepsilon_{kl} + E_{nb}(K \cdot\cdot L) - \Delta E_a^*(K^\circ) - \Delta E_a^*(L^\circ) \tag{12.7}$$

Equation (12.7) is exact.

It is important in another way—it is the key for the description of bond dissociation energies.

Indeed, using here the definition of D_{kl} [Eq. (12.1)] and also that of reorganizational energy [Eq. (12.5)], we deduce from (12.7) that

$$D_{kl} = \varepsilon_{kl} - E_{nb}(K \cdot\cdot L) + \text{CNE} + \text{RE}(K) + \text{RE}(L) \tag{12.8}$$

This energy formula [234,235] is general and suffers from no approximations in that all the appropriate bonded and nonbonded contributions are formally accommodated and because the electroneutrality requirements are met in the definition of reorganizational energy. Only dissociations yielding electroneutral products need be considered as the formation of ions, $KL \rightarrow K^+ + L^-$, require no more than final corrections involving the ionization potential of K and the electron affinity of L.

The formula for D_{kl} contains energy terms, namely, ε_{kl} and $E_{nb}(K \cdot\cdot L)$, which depend only on the properties of the reactant KL. The calculation of CNE, however, requires additional information for use in Eq. (12.4). Finally, this formula also contains information about the products, which is included in the reorganizational energy terms. In other words, there is no way that dissociation energies could be predicted exclusively in terms of the reactant's groud-state properties. Although unfortunate, this point must be clear in our minds: Eq. (12.8) is only capable of telling why D_{kl} and ε_{kl} are different. Yet again, this is instructive, because all the terms appearing in (12.8) are understood on physical grounds.

The true merits of this equation are revealed by a survey of the leading terms governing bond dissociations when the cleavage occurs (1) in the "interior" of the molecule (i.e., when both K and L are polyatomic groups), (2) in its peripheral region [i.e., when K (or L) is an atom], or (3) when it concerns "exterior" bonds formed by a molecule, such as hydrogen bonds. These topics shall be discussed shortly.

The application of (12.8) to diatomic molecules merits a few words, to help develop familiarity with this equation. Evidently, $E_{nb} = 0$ for these molecules. Eq. (12.5) gives $\text{RE}(K) = \Delta E_a^*(K^\circ)$ because $\Delta E_a^*(K\cdot) = \Delta E_a^*(\text{atom } K) = 0$. Equation (12.2) becomes $\Delta E_a^*(KL) = \varepsilon_{kl}$. Using it in Eq. (12.7), we get $\text{CNE} + \text{RE}(K) + \text{RE}(L) = 0$ for any diatomic molecule. Finally, Eq. (12.8) yields the well-known result

$D_{kl} = \varepsilon_{kl}$. On the other hand, Eq. (12.4) shows that $\Delta E_a^*(K^\circ) = 0$, meaning that $RE(K) = 0$ when K is an atom. This means that $CNE = 0$ for diatomic molecules. Now, CNE accounts for the fact that any charge imbalance affects the chemical bonds and the nonbonded terms making up the energies of the subunits K and L. For that reason it concerns only polyatomic subunits. In writing $CNE = 0$ for heteronuclear diatomic molecules, it is understood that the charge neutralization accompanying their cleavage is part of ε_{kl}.

12.3   NONBONDED INTERACTIONS

Although small, the nonbonded interactions $E_{nb}(K \cdots L)$ play a role in accurate applications of Eq. (12.8). Selected results are presented in Table 12.2, for future use.

The approach is that adopted earlier for E_{nb} [Eq. (10.3)], using the point-charge model, which seems reasonable for describing interactions between distant atoms, at least in sigma systems. The charges and the geometries are the same as those used earlier for the total nonbonded interactions represented by Eq. (10.3).

Calculations of this sort are tedious, but still require some attention. The reward is found in the fact that we are now satisfied with the idea that these terms are, indeed, very small and unlikely to distort in any way the overall picture offered by this theory.

TABLE 12.2. Nonbonded Interactions (kcal/mol)

R_1	R_2	$E_{nb}(R_1 \cdots R_2)$	$E_{nb}(R_1 \cdots H)$
CH_3	CH_3	-0.225	0.046
CH_3	C_2H_5	-0.196	—
CH_3	$n\text{-}C_3H_7$	-0.189	—
CH_3	$i\text{-}C_3H_7$	-0.175	—
CH_3	$n\text{-}C_4H_9$	-0.196	—
CH_3	$i\text{-}C_4H_9$	-0.181	—
CH_3	$s\text{-}C_4H_9$	-0.171	—
CH_3	$t\text{-}C_4H_9$	-0.164	—
C_2H_5	C_2H_5	-0.167	0.039
C_2H_5	$n\text{-}C_3H_7$	-0.159	—
C_2H_5	$i\text{-}C_3H_7$	-0.133	—
$n\text{-}C_3H_7$	$n\text{-}C_3H_7$	-0.152	0.036
$i\text{-}C_3H_7$	$i\text{-}C_3H_7$	-0.107	0.018
$n\text{-}C_4H_9$	$n\text{-}C_4H_9$	-0.167	0.038
$i\text{-}C_4H_9$	$i\text{-}C_4H_9$	-0.138	0.028
$s\text{-}C_4H_9$	$s\text{-}C_4H_9$	-0.116	0.008
$t\text{-}C_4H_9$	$t\text{-}C_4H_9$	-0.06	-0.019

The atoms are positioned as in the parent hydrocarbons, using experimental or calculated (STO-3G) geometries. The charges correspond to the reference atomic charge, $q_c = 35.1$ me for ethane.

Source: Reproduced from Ref. 234.

12.4 SELECTED REORGANIZATIONAL ENERGIES

The heats of formation of alkyl radicals are comprehensively reviewed [236]. In our calculations [234], two sets were considered: those of Castelhano and Griller [237] and a set of $\Delta H_f(R\cdot)$ values given in a critical selection [238]. Although some measurements were carried out in solution, they can be regarded as being equivalent to gas-phase data to a good approximation, for reasons given in detail elsewhere [237,239]. Our "balanced" selection of $\Delta H_f(R\cdot)$ values [234] is supported by more recent experimental data [237,240] and a critical review of earlier work [241]. The atomization energies of these radicals, $\Delta E_a^*(K\cdot)$, reported in Table 12.3 were deduced from Eq. (9.6), valid for nonlinear molecules. The reorganizational energies were then calculated by means of Eq. (12.5), using the appropriate $\Delta E_a^*(K°)$ energies reported in Ref. 234. The zero-point+heat content energies required for solving Eq. (9.6) are indicated in Chapter 9.

[The $\Delta E_a^*(K\cdot)$ and the corresponding RE values are indicated with a precision that is not warranted by our actual knowledge. They are used as indicated primarily in order to facilitate recalculations without being continually bothered by roundoff errors.]

The easiest route to RE is by means of Eq. (12.5), provided, of course, that we have access to both $\Delta E_a^*(K°)$ and $\Delta E_a^*(K\cdot)$. Unfortunately, this is not often the case. Example 12.1 presents such a simple situation. More involved situations will be considered at a later stage.

TABLE 12.3. Radical Atomization and Reorganizational Energies (kcal/mol)

Radical K·		$\Delta E_a^*(K\cdot)$	RE[a]
1	$CH_3\cdot$	307.89	12.45
2	$C_2H_5\cdot$	602.73	10.05
3	$n\text{-}C_3H_7\cdot$	896.88	9.19
4	$iso\text{-}C_3H_7\cdot$	899.68	6.99
5	$n\text{-}C_4H_9\cdot$	1191.05	8.94
6	$iso\text{-}C_4H_9\cdot$	1193.26	7.62
7	$sec\text{-}C_4H_9\cdot$	1194.85	4.71
8	$tert\text{-}C_4H_9\cdot$	1199.16	2.45
9	$cyclo\text{-}C_6H_{11}\cdot$	1657.22	5.81
10	$CH_2{=}CH\cdot$	450.40	9.53
11	$CH_2\cdot$	190.85	20.57
12	$NH_2\cdot$	181.06	19.10
13	$CH_3NH\cdot$	471.54	9.19
14	$NH_2CH_2\cdot$	478.00	4.70

Items **1–9** are from Refs. 44, 234, and 235. Items **9–14** are explained in the text. (Chapter 15 explains the theoretical results obtained for **12** and **13**.)

Example 12.1. For the CH_2=CH· radical, consider the central C_2—C_3 bond of 1,3-butadiene and its bond energy, $\varepsilon_{CC} = 90.25$ kcal/mol ($q_C = 9.86$ me). With ΔE_a^*(butadiene) $= 1010.1$ kcal/mol (from $\Delta H_f^\circ = 26.33$ kcal/mol [242]) it is thus $\Delta E_a^*(K^\circ) \simeq 459.93$ kcal/mol for the electroneutral CH_2=CH group while it still is part of butadiene. Finally, knowing that $\Delta E_a^*(K^\bullet) = 450.40$ kcal/mol for the radical ($\Delta H_f^\circ = 63.45$ kcal/mol [203] and ZPE $+ H_T - H_0 = 23.20$ kcal/mol [205]), we get RE(CH_2=CH·) $= 9.53$ kcal/mol.

Likewise, RE(CH_2) follows from the energy of CH_2, $\Delta E_a^*(CH_2) = 190.85$ kcal/mol (from $\Delta H_f^\circ = 92.35$ kcal/mol [203] and ZPE $+ H_T - H_0 = 12.58$ kcal/mol [203]. The C=C bond energy of ethylene, $\varepsilon_{CC} = 139.37$ kcal/mol (Table 12.2), and its atomization energy, $\Delta E_a^*(C_2H_4) = 562.22$ kcal/mol, Chapter 14, give $\Delta E_a^*(K^\circ) = 211.42$ kcal/mol and thus the RE of Table 12.3.

12.5 APPLICATIONS

The hydrocarbons are good candidates for highlighting the merits of our master equation (12.8), its salient features and some intricacies plaguing the interpretation of trends observed for the breaking of chemical bonds. The use of Eq. (12.8) requires the appropriate RE energies (Table 12.3) as well as the charge neutralization energies, CNE [Eq. (12.7)].

Now we come to the interesting part: the skeleton of carbon–carbon bonds, which represents a typical example of "interior bonds."

The Carbon Skeleton of Alkanes

The first thing to do is to calculate the charge neutralization energy, CNE, by means of Eq. (12.7). The appropriate $\Delta E_a^*(KL)$ results (which are described in Chapter 13) are indicated in Table 12.5 along with the corresponding CC bond energies.

The nonbonded contributions are those of Table 12.2. The $\Delta E_a^*(K^\circ)$'s are displayed in Table 12.4.

TABLE 12.4. Calculation of $\Delta E_a^*(K^\circ)$ (kcal/mol)

K		$\Delta E_a^*(KK)$	ε_{CC}	$\Delta E_a^*(K^\circ)$
1	CH_3	710.54	69.63	320.34
2	C_2H_5	1298.10	72.38	612.78
3	n-C_3H_7	1885.69	73.39	906.07
4	iso-C_3H_7	1887.16	73.71	906.67
5	n-C_4H_9	2473.21	73.06	1199.99
6	iso-C_4H_9	2476.03	74.14	1200.88
7	sec-C_4H_9	2473.55	74.44	1199.50
8	$tert$-C_4H_9	2477.09	73.87	1201.61
9	$cyclo$-C_6H_{11}	3401.42	75.27	1663.03
10	CH_2=CH	1010.1	90.25	459.93

The important point is that the CNE contributions are small in comparison with the significant changes of the ε_{CC} bond energies. This was to be expected because charge neutralization between alkyl groups benefits from important compensations. In simple words, what one group looses when dissociation occurs is largely recovered by its partner because both groups consist of similar CC and CH bonds.

Nonzero CNEs should be generally expected when $K \neq L$, but their smallness and the almost negligible variations of the nonbonded part suggests that $CNE - E_{nb}(K \cdot \cdot L) \simeq$ constant should represent a valid approximation, that is

$$D_{CC} \simeq \varepsilon_{CC} + RE(K) + RE(L) + 0.33 \, \text{kcal/mol} \tag{12.9}$$

where the constant (0.33) represents the average of the CNE plus the nonbonded contributions evaluated for 36 molecules constructed from the eight alkyl groups considered here. The error made in this manner (standard deviation = 0.26 kcal/mol) is certainly acceptable.[3]

Table 12.5 lists the results obtained with this approximation. [There is no point in showing those given by Eq. (12.8) as they duplicate the experimental results deduced from Eq. (12.1) with the $\Delta E_a^*(K \cdot)$ data of Table 12.3 and the appropriate $\Delta E_a^*(KL)$ data.]

The approximate validity of Eq. (12.9) highlights the leading role played by the reorganizational energies but also vividly demonstrates the importance of using the correct intrinsic bond energies, whose values range from 69.63 to $\sim$89.1 kcal/mol in these examples.

The merit of Eqs. (12.8) and (12.9) is that they allow us to probe into individual chemical bonds and their intrinsic energies ε_{kl}—which cannot be observed experimentally—through their link with the corresponding dissociation energies that can be measured. In other words, beyond knowing that the ε_{kl} energies add up correctly, we can also check that the right energy is associated with the right bond.

The link between dissociation and intrinsic bond energies evidently transcends academic curiosity because of its important down-to-earth practical aspects: chemistry, after all, is an exercise of making and breaking chemical bonds, more often than not influenced by the environment in which the reactions occur. ε_{kl} is observation, D_{kl} is action.

This brings us to examine the formal resemblance and the conceptual difference between Eq. (12.9) and Sanderson's approximation [244,245]. The latter can be deduced from Eqs. (12.1) and (12.3) by redefining the reorganizational energies as $\Delta E_a^*(K) - \Delta E_a^*(K \cdot)$—an approach oblivious of the fact that K and K· are not isoelectronic in most cases, but approximately valid as long as the CNE terms remain sufficiently small. The point is that Sanderson's formula, $D_{CC} = \varepsilon_{CC} + RE(K) + RE(L)$, treats ε_{CC} as a constant, which is not true. This simplifying hypothesis did not appear

---

[3]This conclusion also holds for dissociation enthalpies $DH(KL) = \Delta H_f(K \cdot) + \Delta H_f(L \cdot) - \Delta H_f(KL)$, where the ΔH_f terms are the appropriate enthalpies of formation. This is due to structure-related regularities characterizing zero-point and heat content energies. Equation (12.9) thus gives $DH_{CC} \simeq \varepsilon_{CC} + RE(K) + RE(L) - 5.9 \, \text{kcal/mol}$.

TABLE 12.5. Dissociation of Selected CC Bonds (kcal/mol)

Bond	ΔE_a^*(KL)	ε_{CC}	CNE	$D_{CC}{}^a$ Eq. (12.9)	Experimental
CH_3—CH_3	710.54	69.63	0.00	94.86	94.76
CH_3—C_2H_5	1004.29	71.14	−0.16	93.97	93.67
CH_3—nC_3H_7	1298.10	71.56	−0.06	93.53	93.33
CH_3—iC_3H_7	1300.02	72.42	0.42	92.19	92.45
CH_3—nC_4H_9	1591.90	71.41	−0.04	93.13	92.96
CH_3—iC_4H_9	1593.17	71.91	−0.13	92.31	92.02
CH_3—sC_4H_9	1593.17	72.53	0.63	90.02	90.43
CH_3—tC_4H_9	1596.06	73.14	0.81	88.37	89.01
C_2H_5—C_2H_5	1298.10	72.38	0.00	92.81	92.64
C_2H_5—nC_3H_7	1591.90	72.89	0.00	92.46	92.29
C_2H_5—iC_3H_7	1593.17	73.23	0.36	90.60	90.76
C_2H_5—nC_4H_9	1885.69	72.74	0.01	92.06	91.91
C_2H_5—iC_4H_9	1887.00	73.23	−0.03	91.23	91.01
C_2H_5—sC_4H_9	1886.33	73.55	0.36	88.64	88.75
C_2H_5—tC_4H_9	1888.41	73.63	0.31	86.46	86.52
nC_3H_7—nC_3H_7	1885.75	73.40	0.00	92.11	91.93
nC_3H_7—iC_3H_7	1887.00	73.81	0.32	90.32	90.44
nC_3H_7—sC_4H_9	2180.15	74.10	0.34	88.33	88.42
iC_3H_7—iC_3H_7	1887.16	73.71	0.00	88.02	87.80
iC_3H_7—sC_4H_9	2180.32	74.05	−0.01	86.08	85.79
iC_3H_7—tC_4H_9	2182.36	73.88	0.15	83.65	83.52
iC_4H_9—sC_4H_9	2475.21	74.49	0.22	87.15	87.10
sC_4H_9—sC_4H_9	2473.55	74.44	0.00	84.19	83.85
sC_4H_9—tC_4H_9	2476.04	74.47	0.40	81.96	82.03
tC_4H_9—tC_4H_9	2477.09	73.87	0.00	79.10	78.77
cC_6H_{11}—CH_3	2056.94	72.96	—	91.55	91.83
cC_6H_{11}—C_2H_5	2350.24	73.95	—	90.14	90.29
cC_6H_{11}—nC_4H_9	2937.50	74.27	—	89.35	89.23
cC_6H_{11}—tC_4H_9	2939.05	74.67	—	83.26	82.67
cC_6H_{11}—cC_6H_{11}	3401.43	75.27	0.00	87.22	86.99
NH_2CH_2—CH_3	875.36	72.83	—	90.31	89.47
NH_2CH_2—C_2H_5	1169.62	73.99	—	89.07	88.89
NH_2CH_2—nC_3H_7	1463.68	74.51	—	88.73	88.80
NH_2CH_2—CH_2NH_2	1041.54	75.41	0.00	85.14	85.54
CH_2:CH—CH:CH_2	1010.1	90.25	0.00	109.64	109.30

The experimental results were deduced from Eq. (12.1). Those given under the heading Eq. (12.9) were calculated using carbon atomic charges deduced from ^{13}C NMR shifts reported in Ref. 243 (for C_2H_6), in Ref. 169 (for the C_3H_8—C_6H_{14} hydrocarbons) and in Ref. 166 for the larger molecules. For the amines, see Chapter 15.

TABLE 12.6. Dissociation and Intrinsic Energies of Selected CH Bonds (kcal/mol)

CH Bond	D_{CH}	ε_{CH}
CH_3—H	111.42	104.86
C_2H_5—H	107.81	106.81
n-C_3H_7—H	107.41	107.13
i-C_3H_7—H	104.61	108.72
n-C_4H_9—H	107.05	107.23
i-C_4H_9—H	106.76	107.59
s-C_4H_9—H	103.25	109.35
t-C_4H_9—H	100.86	110.89

as a conceptual stumbling block at that time—and, indeed, many interesting results were obtained in this fashion—but, as we now know, any scheme postulating constant bond energies conflicts in principle with the basics demanding conservation of molecular electroneutrality. The quality of the theoretical intrinsic bond energies is clearly reflected in the quality of the bond dissociation energies given by Eq. (12.9).

Carbon–Hydrogen Bonds

The carbon–hydrogen bonds are the prototype of "peripheral bonds." In the perspective of the CNE effect, the extraction of an atom that was partially charged in the host

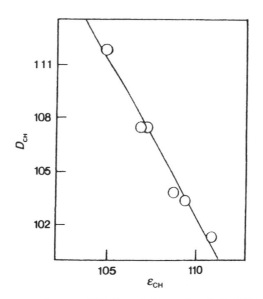

Figure 12.1. A comparison between CH dissociation and intrinsic CH bond energies (kcal/mol) [Eq. (12.10)] [233].

molecule affects only the bond energies of the polyatomic fragment left behind because the extracted atom forms no bonds and cannot compensate for energy changes induced in the other fragment. Briefly, any cleavage of this sort is expected to involve a significant charge neutralization energy. CH bonds fall in with this description. But the outcome is surprising (Table 12.6).

The comparison between dissociation and intrinsic bond energies is intriguing, to say the least. It suggests (Fig. 12.1) that those bonds that contribute more to thermochemical stability are the ones that break up more easily. This is certainly a point requiring clarification:

$$D_{CH} \simeq 295.91 - 1.76\,\varepsilon_{CH} \qquad (12.10)$$

Sanderson's claim that all contributing CH bond energies are equal also merits scrutiny, although it contradicts the very basics of our description of charge-dependent bond energies satisfying charge normalization.

It turns out that the CNE part accompanying CH dissociations not only is quantitatively important but also produces a remarkable effect (Table 12.7). Indeed, CNE compensates the major part, if not all, of the differences existing between ε_{CH} energies: $\varepsilon_{CH} + $ CNE is nearly constant for all the $C(sp^3)$—H bonds, on one hand, and also for the $C(sp^2)$—H bonds, on the other hand. Hence, incorporating now the small (<0.05 kcal/mol) nonbonded interactions into this approximation, we rewrite Eq. (12.8) as follows

$$D_{CH} \simeq \text{constant} + \text{RE(K)} \qquad (12.11)$$

TABLE 12.7. CNE Energies for $C(sp^3)$—H and $C(sp^2)$—H bond dissociations (kcal/mol)

Bond	ε_{CH}	CNE	$\varepsilon_{CH} + $ CNE
CH_3—H	104.86	-5.81	99.05
C_2H_5—H	106.81	-9.01	97.80
n-C_3H_7—H	107.42	-8.87	98.55
i-C_3H_7—H	108.71	-11.08	97.63
n-C_4H_9—H	107.49	-9.08	98.41
i-C_4H_9—H	107.98	-8.42	99.56
s-C_4H_9—H	109.35	-10.74	98.61
t-C_4H_9—H	110.86	-12.50	98.36
c-C_6H_{11}—H	110.28	-11.50	98.78
$CH_2{:}CHCH_2$—H	107.3	-8.6	98.69
$CH_2{:}CHCH_2CH_2$—H	107.4	-8.9	98.52
$C_6H_5CH_2$—H	108.0	-9.0	99.0
$CH_2{:}CH$—H	105.72	-3.11	102.61
$CH_3CH{:}CH$—H	106.0	-3.9	102.1
$(CH_3)_2C{:}CH$—H	104.6	-1.3	103.3
$CH_2{:}C(CH_3)$—H	109.9	-7.1	102.8
C_6H_5—H	112.36	-9.79	102.57

where the constant, $\sim$98.40 or 102.65 kcal/mol for $C(sp^3)$—H and $C(sp^2)$—H bonds, respectively, indicate the appropriate average value of $\varepsilon_{CH} + CNE - E_{nb}(K \cdot \cdot H)$.

The ε_{CH} energies are genuine CH bond energies. CNE disguises them in such a way that, when viewed from the perspective of dissociation energies, all CH bonds involving the same type of carbon are perceived as if they were equal in energy, to a good approximation.

This echoes Sanderson's claim [244,245] that all contributing CH bond energies are equal. The behavior observed here for the CH bonds is typical for peripheral bonds. Indeed, it holds true also for CX bonds (X = Cl,Br,I), where $\varepsilon_{CX} + CNE \simeq$ constant [44].

The approximate linear decrease of reorganizational energies with increasingly larger ε_{CH} energies (Fig. 12.2) and Eq. (12.11) explain the existence of a correlation between D_{CH} and ε_{CH}.

Applications

The approximations (12.10) and (12.11) can be used to obtain information about reorganizational energies, as shown in the following two examples.

Example 12.2. Here we deduce the RE of the $NH_2CH_2 \cdot$ radical.

Method 1. This method follows from Eq. (12.5). For $NH_2CH_2 \cdot$ we use $\Delta E_a^* = 478.00$ kcal/mol, corresponding to $\Delta H_f^\circ = 38 \pm 2$ kcal/mol [246] and ZPE + $H_T - H_0 = 32.62$ kcal/mol [44]. For NH_2CH_2—CH_2NH_2, on the other hand, we find $\Delta E_a^* = 1041.54$ kcal/mol (from $\Delta H_f^\circ = -4.07 \pm 0.14$ kcal/mol [247])

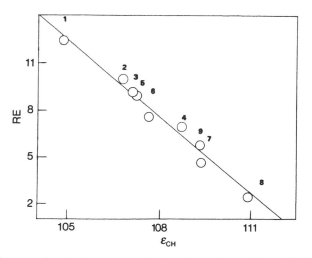

Figure 12.2. A comparison between reorganizational energies and theoretical bond energies of alkane CH bonds (kcal/mol). The numbering refers to that shown in Table 12.3 [233].

and ZPE $+ H_T - H_0 = 70.80$ kcal/mol [34]. For the central CC bond of this molecule, it is $\varepsilon_{CC} = 75.41$ kcal/mol (with $q_C = 29.18$ me, from its ^{13}C NMR spectrum [165]). Assuming $E_{nb} \simeq -0.23$ kcal/mol, Eq. (12.4) gives $\Delta E_a^*(K^\circ) \simeq 482.95$ kcal/mol and thus RE $\simeq 4.95$ kcal/mol, from Eq. (12.5).

Method 2. For CH_3NH_2 we use $\Delta E_a^* = 580.95$ kcal/mol (deduced from $\Delta H_f^\circ = -5.50 \pm 0.12$ kcal/mol [248] and ZPE $+ H_T - H_0 = 41.48$ kcal/mol, from the spectra of Ref. 249). For $NH_2CH_2\cdot$ we just found $\Delta E_a^* = 478.00$ kcal/mol. Hence $D_{CH} = 102.95$ kcal/mol, from Eq. (12.1). Knowing that $\varepsilon_{CH} + CNE$ range from ~ 98.0 to ~ 98.6 kcal/mol in Eq. (12.11), one can estimate $RE(NH_2CH_2\cdot)$ at ~ 4.70 kcal/mol.

Example 12.3. For 1,1'-bicyclohexyl, $\Delta E_a^*(KK) = 3401.43$ and $\Delta E_a^*(K\cdot) = 1657.22$ kcal/mol; $\varepsilon_{CC} = 75.27$ and $E_{nb} = -0.1$ kcal/mol give RE $= 5.81$ kcal/mol. With $\varepsilon_{CH} = 109.28$ kcal/mol for cyclohexane, Eq. (12.10) gives $D_{CH} \simeq 103.6$ kcal/mol; thus RE $\simeq 5.3$ kcal/mol from (12.11) [234].

Carbon–Nitrogen Bonds

A test resembling those given for the alkanes is offered in Table 12.8 for CN bonds.

Straightforward general bond energy theory is applied in the calculation of intrinsic CN bond energies by means of the equation

$$\varepsilon_{CN} = \varepsilon_{CN}^\circ + a_{CN}\Delta q_C + a_{NC}\Delta q_N \tag{12.12}$$
$$= 60.44 - 0.603 \times \Delta q_C - 0.448 \times \Delta q_N \tag{12.13}$$

with charges deduced from the appropriate carbon–13 and nitrogen–15 chemical shifts. The corresponding dissociation energies follow from Sanderson's approximation [244,245]

$$D_{kl} \simeq \varepsilon_{kl} + RE(K) + RE(L) \tag{12.14}$$

and the use of the reorganizational energies of Table 12.3. These results are in satisfactory agreement with their counterparts given by Eq. (12.1) and the relevant thermochemical data, namely, the atomization energies deduced from experimental enthalpies of formation and the relevant ZPE $+ H_T - H_0$ energies.

The relatively modest changes observed for D_{CN} in this series should not obscure the fact that the corresponding intrinsic bond energies (ε_{CN}) cover a range of ~ 10 kcal/mol. It so happens that when a bond energy increases, the reorganizational energy of the corresponding alkyl radical diminishes, one effect largely compensating for the other.

Similar conclusions hold for nitroalkanes as well; they were calculated by means of Eq. (12.12), with $\varepsilon_{CN}^\circ = 53.00$ kcal/mol. The NMR shifts reported in Ref. 138 were used for the carbons; those of nitrogen are from [156]. The conclusions are in no way affected by the simplifying assumption RE $= 0$ for NO_2, which may well be revised in the future.

TABLE 12.8. CN Bond Dissociation Energies of Selected Alkylamines and Nitroalkanes (kcal/mol)

Molecule	Δq_N	Δq_C	ε_{CN}	D_{CN} Calculated	D_{CN} Experimental
CH_3NH_2	0	0	60.44	91.99	92.00
$C_2H_5NH_2$	-4.72	-1.27	63.32	92.47	91.57
n-$C_3H_7NH_2$	-3.81	-2.41	63.60	91.89	91.68
iso-$C_3H_7NH_2$	-8.55	-2.17	65.58	91.67	91.70
n-$C_4H_9NH_2$	-3.90	-2.08	63.44	91.48	91.57
iso-$C_4H_9NH_2$	-3.18	-3.30	63.86	90.58	90.62
sec-$C_4H_9NH_2$	-7.65	-3.03	65.70	89.45	89.97
$tert$-$C_4H_9NH_2$	-11.55	-2.80	67.30	88.85	89.74
$cyclo$-$C_6H_{11}NH_2$	-8.04	-3.36^a	66.07	90.98	91.50
$(CH_3)_2NH$	-14.35	-1.33^b	67.67	89.31	89.27
$(CH_3)(C_2H_5)NH$	-18.67	-2.52^b	70.32^c	89.56	89.32
CH_3NO_2	0	0	53.00^d	65.45	65.05
$C_2H_5NO_2$	-3.03	-1.94	55.53	65.58	65.58
iso-$C_3H_7NO_2$	-6.07	-3.23	57.67	64.66	65.25

Calculated from the ^{13}C shift of methylcyclohexane [250], following the rules given in Ref. 165 to obtain the shift of the α-carbon of the cyclohexyl ring.
From the ^{13}C shift of the parent alkane [169] and the rules given in Ref. 165.
This result is for the bond formed with the ethyl group. With $\Delta q_C = -0.95$ me for the CH_3 carbon, it is $\varepsilon_{CN} = 69.38$ kcal/mol for the CH_3—NH bond.
RE(NO_2) is expected to be small, judging by the smallness of the dissociation energy of O_2N—NO_2 (~ 16 kcal/mol), and has been included in this entry.

The following examples also illustrate the insight one can gain into the way bond dissociation D_{kl} reacts on modifications affecting ε_{kl}.

Example 12.4: Influence of the Environment on D_{kl}. Nitromethane is interesting to some people because it explodes. The reason is, of course, in the cleavage of the carbon–nitrogen bond. The monomer, compared to its trimer (taken as a model for the crystal), reveals that the C and N net charges change by $\Delta q_C \simeq -8.7$ and $\Delta q_N \simeq 1.1$ me, respectively, on "crystallization." Our bond energy formula and the appropriate a_{kl} parameters thus indicate that the crystalline environment reinforces the CN bond by ~ 4.7 kcal/mol, which is significant at the *local* point of rupture, responsible for the reaction [251].

A more attractive example is offered by the $C_{1'}$—N_9 bonds of deoxyadenosine and deoxyguanosine and their response to electrophilic attacks on the purinic N atoms [252]. Cleavage of these bonds leads to depurination, and its physics is relevant in the understanding of the depurination of DNA structures by radical cations. Among other results, in terms of the attack on N_7 of the adenine or guanine base, applications of our bond energy methods point to a significant weakening of the glycosyl CN bond. Now, most of the substituents whose attack on N_7 has been studied [253] were found to provoke rapid cleavage of the CN bond, resulting in instantaneous depurination. It seems reasonable to argue that at least part of the explanation addressing the depurination of DNA has something to do with the bond weakening due to local charge effects. It is true that electrophiles like H^+, CH_3^+, ... come to stick on the purinic base, thus lowering its RE and D_{CN} by the same small amount, but this still leaves us with $\Delta D_{CN} \simeq \Delta \varepsilon_{CN}$ as a reasonable first estimate. Any lowering of RE (and of D_{CN}) would indeed only add to the effect triggered by the thinning of the electronic charge at the bond-forming atoms.

These examples are rough estimates, of course, with no pretention to rigor, but they illustrate vividly the merits of Eq. (12.8). Through this formula, a solid knowledge of modifications suffered by the charges of individual bonds as a result of external influences has the potential of guiding the interpretation of bond ruptures enhanced (or triggered) by events at a distance from the place where the reaction takes place—or explain the opposite response if that should be the case.

Equation (12.8) offers promising new fields of applications regarding the dissociation of chemical bonds.

12.6   CONCLUSION

The formula describing bond dissociation [Eq. (12.8)] establishes an exact relationship between dissociation and intrinsic bond energies. The former are amenable to experimental (thermochemical) measurements and thus validate the intrinsic bond energies given by Eq. (10.37) and their one-to-one correspondence with them; in other words, appropriate experimental verifications confirm that the right energy is associated with the right bond and that our bond energy formula (10.37) does, indeed, correctly describe the intrinsic energies of chemical bonds.

Conversely, a consideration of possible changes affecting intrinsic bond energies due to modifications of electron densities induced by the environment of a molecule suggests—with the help of Eq. (12.8)—how this environment can promote or retard the dissociation of one or another bond of particular interest in that molecule. This outlook hints at a rich potential of future research exploiting charge analyses to gain a fresh insight into *local* molecular properties, say, into bond energies, first, and, going from there, into matters of great import regarding the making and breaking of chemical bonds.

PART III

APPLICATIONS

CHAPTER 13

SATURATED HYDROCARBONS

13.1 ACYCLIC ALKANES

The comparison between theoretical and experimental results presented here for acyclic alkanes is made with the help of Eq. (10.50), repeated here for convenience (kcal/mol):

$$\Delta E_a^* \simeq 710.54 + 290.812(n-2) + 0.03244 \sum_k N_{C_kC}\delta_{C_k} + 0.05728 \sum_k \delta_{C_k}$$

This approximation includes the small nonbonded term and is most handy because it requires only the appropriate set of NMR shifts δ_{C_k} (ppm) from ethane. The loss in precision in very small, with respect to calculations including explicit nonbonded terms, but the economy in effort is major. Remember that N_{C_kC} is the number of CC bonds formed by the C atom whose ^{13}C shift is δ_{C_k}. The formula (6.8)

$$q_C = 35.1 - 0.148\,\delta_C \text{ (me)}$$

converts the δ_C variable into its equivalent charge result.

The present selection of experimental ΔH_f° values is not intended to represent a critical collection of "best values." Slightly different results are reported for some molecules in Refs. 255–257, but experimental uncertainties of this magnitude do not affect the general conclusions drawn in this work. Regarding our calculations,

---

TABLE 13.1. Standard Enthalpy of Formation of Alkanes (kcal/mol)

	Molecule	ΔE_a^*	ZPE + ΔH	ΔH_f° Calculated	ΔH_f° Experimental[a]
1	Methane	419.27	26.69	−17.73	−17.89 ± 0.07
2	Ethane	710.54	47.91	−20.15	−20.24 ± 0.12
3	Propane	1004.34	66.12	−25.11	−25.02 ± 0.12
4	n-Butane	1298.14	84.33	−30.07	−30.03 ± 0.16
5	Isobutane	1300.13	83.99	−32.40	−32.42 ± 0.13
6	n-Pentane	1591.95	102.54	−35.05	−35.16 ± 0.24
7	Isopentane	1593.24	102.20	−36.68	−36.73 ± 0.14
8	2,2-Dimethylpropane	1596.28	101.86	−40.07	−40.14 ± 0.15
9	n-Hexane	1885.75	120.76	−39.98	−39.96 ± 0.19
10	2-Methylpentane	1887.11	120.41	−41.72	−41.66 ± 0.25
11	3-Methylpentane	1886.38	120.41	−40.99	−41.02 ± 0.23
12	2,2-Dimethylbutane	1888.66	120.07	−43.61	−44.35 ± 0.23
13	2,3-Dimethylbutane	1887.38	120.07	−42.33	−42.49 ± 0.24
14	n-Heptane	2179.28	138.97	−44.85	−44.89 ± 0.22
15	2-Methylhexane	2180.87	138.63	−46.63	−46.60 ± 0.30
16	3-Methylhexane	2180.15	138.63	−45.91	−45.96 ± 0.30
17	3-Ethylpentane	2178.79	138.63	−44.55	−45.29 ± 0.32
18	2,2-Dimethylpentane	2183.10	138.28	−49.21	−49.20 ± 0.37
19	2,3-Dimethylpentane	2180.38	138.28	−46.50	−47.33 ± 0.20
20	2,4-Dimethylpentane	2182.17	138.28	−48.28	−48.21 ± 0.29
21	2,2,3-Trimethylbutane	2182.36	137.94	−48.81	−48.87 ± 0.33
22	n-Octane	2473.22	157.18	−49.81	−49.86 ± 0.25
23	2-Methylheptane	2474.68	156.84	−51.61	−51.47 ± 0.36
24	3-Methylheptane	2474.15	156.84	−51.08	−50.79 ± 0.33
25	4-Methylheptane	2474.05	156.84	−50.98	−50.66 ± 0.33
26	2,2-Dimethylhexane	2476.53	156.50	−53.80	−53.68 ± 0.32
27	2,3-Dimethylhexane	2474.25	156.50	−51.52	−51.10 ± 0.40
28	2,4-Dimethylhexane	2475.21	156.50	−52.48	−52.40 ± 0.33
29	2,5-Dimethylhexane	2476.05	156.50	−53.32	−53.18 ± 0.40
30	3,3-Dimethylhexane	2474.87	156.50	−52.14	−52.58 ± 0.32
31	3,4-Dimethylhexane	2473.57	156.50	−50.84	−50.87 ± 0.40
32	2,2,3-Trimethylpentane	2475.85	156.15	−53.47	−52.58 ± 0.40
33	2,2,4-Trimethylpentane	2476.72	156.15	−54.34	−53.54 ± 0.37
34	2,3,3-Trimethylpentane	2474.11	156.15	−51.73	−51.69 ± 0.38
35	2,3,4-Trimethylpentane	2474.45	156.15	−52.07	−51.94 ± 0.43
36	n-Nonane	2766.97	175.40	−54.71	−54.66 ± 0.25
37	4-Methyloctane	2767.76	175.05	−55.85	−56.19
38	235-Trimethlhexane	2769.11	174.37	−57.88	−57.97

[a]The experimental values of **1–16** are reported in Ref. 247. The results for **37** and **38** are from Ref. 254. All other results are given in Ref. 248.

they were systematically carried out with maximum precision, so as to avoid irritating roundoff errors.

The results are reported in Table 13.1. The zero-point plus heat content energies, abbreviated as ZPE + ΔH, are those described in Chapter 9, Eq. (9.9). The ^{13}C shifts are from Refs. 166, 169, and 243.

The fine-tuning of molecular ΔE_a^* energies, responsible for the differences between structural isomers, rests entirely with small but extremely important modifications of charge distributions affecting $\sum_k \sum_l a_{kl} \Delta q_k$. Because the sum of the $\sum_l a_{kl}$ values is always larger for sp^3 carbons than for hydrogen, it follows that in comparisons between isomers or conformers, the more stable form is the one with the electron-richest carbon skeleton, reflected by larger (downfield) δ_C values.

13.2   CYCLOALKANES

This section is about alkylcyclohexanes and related polycyclic molecules consisting of chair six-membered rings. Our work must thus accommodate conformational features such as those commonly described as butane-*gauche* interactions.

The latter are, indeed, of considerable interest. They have a long history in conformational chemistry [258,259] and deserve attention for the major role they play in the discussion and prediction of structural features. Typically, we refer here to *gauche* interactions exemplified by one of the methyl protons of the axial methylcyclohexane (for instance) interacting with the axial protons at C-3 and C-5 of the ring, or to the three *gauche* interactions occurring in *cis*-decalin.[1]

The occurrence of these interactions is not under dispute. The question lies with the interpretation of *gauche* interactions—are they somehow related to the vibrational energy content of the molecule, or should they rather be traced back to a particularity in the chemical binding in the vibrationless state? The answer is given in Chapter 9.

Again we use Eq. (10.50) for our calculations. Here we write it as

$$\Delta E_a^* \simeq 710.54(1 - m) + 290.812(n - 2 + 2m)$$

$$+ 0.03244 \sum_k N_{C_k C} \delta_{C_k} + 0.05728 \sum_k \delta_{C_k} + F \qquad (13.1)$$

where F depends on the number of *gauche* interactions. F is the function defined in Eq. (10.26).

The Gauche Interactions

Butane-*gauche* effects do not manifest themselves in the vibrational part, ZPE + $H_T - H_0$. So we return to the atomization energy, $\Delta E_a^* = \sum_{k<l} \varepsilon_{kl} - E_{nb} + F$, of

---

[1]A description and convenient counting of these interactions is offered in Ref. 250, along with a wealth of useful information, namely, the ^{13}C nuclear magnetic resonance (NMR) spectra that give access to the atomic charges of the carbon atoms.

the molecule at its potential minimum; the *gauche* contributions are necessarily part of ΔE_a^*. The nonbonded part, approximated as Coulomb interactions between net atomic charges, Eq. (10.3) [206], is far too small to play a relevant role in that matter. Attention is on the other two terms, the sum of intrinsic bond energies and F; the latter, (10.26), measures what is due to variations of internuclear distances and to changes of electronic centers of charge. No term for *gauche* interactions is explicit in any of these parts, but the fact is that our energy formula (13.1) does not work with $F = 0$ in situations where *gauche* interactions are postulated.

Numerical evaluations reveal where we stand. Our strategy is simple—we apply Eq. (13.1) using the appropriate ^{13}C NMR shifts reported by Grant and coworkers [250,260,261], then use these ΔE_a^* terms in Eq. (9.6) and estimate the residual F values attributed to *gauche* interaction with the help of experimental enthalpies of formation. In the absence of *gauche* interactions, $F = 0$. Otherwise F is associated with *gauche* effects and is written F_g.

Now, the comparison of isomerides differing by the number of *gauche* interactions, for example, *cis*-1,3-dimethylcyclohexane (no *gauche* interaction) versus *trans*-1,3-dimethylcyclohexane (two interactions), or *trans*-decalin (no interaction) versus *cis*-decalin (three interactions) reveals that $\Delta E_a^* - F_g$ consistently decreases by $\sim$1.9 kcal/mol for one *gauche* interaction (average value), which is about twice the commonly accepted value for *gauche* effects [181].

On the other hand, looking at the enthalpies ΔH_f°, the same comparisons show that one *gauche* interaction reduces the thermochemical stability by only $\sim$1.0 kcal/mol, meaning that part of the loss in binding energy is "recovered" by F_g. These are rough numbers, of course; they will be refined later on. Let us first examine the physical content of F_g.

In line with the basic theory that leads to Eqs. (10.40) and (13.1), the function (11.12)

$$F_{kl} = -\frac{3}{7} Z_k^{\text{eff}} Z_l^{\text{eff}} \left[R_{kl}^{-1} - \left(R_{kl}^{-1} \right)^\circ - \left(\langle r_{kl}^{-1} \rangle - \langle r_{kl}^{-1} \rangle^\circ \right) \right]$$

$$- \frac{3}{7} Z_k^{\text{eff}} q_l \left(\langle r_{kl}^{-1} \rangle - \langle r_{kl}^{-1} \rangle^\circ \right)$$

adequately represents the effect of varying internuclear distances and of changing electronic charge centroids [44]. F_{kl} mirrors, so to speak, how nucleus k "sees" the electrons of atom l. For the problem at hand, we consider CH bonds, where $k = C$ and $l = H$. The reference for alkanes is ethane, selected with $(R_{CH}^{-1})^\circ = \langle r_{CH}^{-1} \rangle^\circ$. The second right-hand-side (RHS) term of Eq. (11.12) is very small because hydrogen net charges, q_l, are quite small. Accordingly, we ought to consider only

$$F_{kl} \approx -\frac{3}{7} Z_k^{\text{eff}} Z_l^{\text{eff}} \left(R_{CH}^{-1} - \langle r_{CH}^{-1} \rangle \right) \qquad (13.2)$$

What matters here is how $R_{CH}^{-1} - \langle r_{CH}^{-1} \rangle$ differs from $(R_{CH}^{-1})^\circ - \langle r_{CH}^{-1} \rangle^\circ$. The interaction under scrutiny involves the charges of three hydogen atoms: that of the axial methyl group overhanging the ring and the axial hydrogens on C-3 and C-5 (taking methylcyclohexane as a model). These are repulsive interactions, of course, but simple Coulomb repulsions $q_r q_s / R_{rs}$ of this kind are already included in E_{nb} and need not be singled out at this point. Here we argue that repulsions between the hydrogen electron clouds affect their shape and thus $\langle r_{CH}^{-1} \rangle$. This change of $\langle r_{CH}^{-1} \rangle$ is certainly small, and we feel presently unable to evaluate it by direct calculations of the relevant charge centroids. But the effect is sizable, as shown by the following example. Consider $R_{CH} = 1.093$ Å, which is close to the theoretical value, and assume that the charge centroid of the hydrogen electron is shifted by 0.001 Å, closer to carbon: $r_{CH} = 1.092$ Å. With 1 bohr $= 0.52917$ Å and 1 hartree $= 627.51$ kcal/mol, we thus obtain $F_{kl} \approx 0.48$ kcal/mol. Now, there are three CH bonds for two *gauche* interactions, so that, in this example, we get a correction of 0.72 kcal/mol for one *gauche* interaction. Although this matter must be held in abeyance until direct evaluations of the relevant charge centroids can be performed, we tentatively submit this model for explaining the nature of F_g—whose very existence is unmistakably revealed by the numerical results—within the strict framework of the theory describing charge-dependent bond energies.

The final results obtained from the study [181] of molecules **3, 5, 7, 9, 11**, and **17** of Table 9.1 indicate that $F_g = 0.822$ kcal/mol for one *gauche* interaction, meaning that we must use

$$F = 0.822 \, n_g \tag{13.3}$$

in the calculation of ΔE_a^*, where n_g is the number of corrections required for *gauche* interactions involving axial protons. (Note that $n_g = 2$ for molecules **3, 4, 7, 8**, and **12** of Table 13.2 and $n_g = 3$ for **28**.)

This closes the topic of *gauche* effects in energy calculations. All we have to do is to use Eq. (13.3) in Eq. (13.1).

At long last, we can calculate ΔE_a^* energies of cycloalkanes by means of Eq. (13.1), regardless of whether butane-*gauche* effects intervene. The results are given in Table 13.2. Enthalpies of formation calculated along these lines, using charges inferred from NMR chemical shifts, agree within 0.24 kcal/mol (root-mean-square deviation) with their experimental counterparts [44].

Note that the molecules selected here cover indiscriminately chair and boat, as well as twist–boat six-membered ring structures. Here again, in comparisons between structural isomers, the form with the electron-richer carbon skeleton is thermochemically favored. In other words, the more stable form is given away by the larger sum, $\sum_k \delta_{C_k}$, of its ^{13}C shifts. In a way, *the hydrogen atoms play the role of a reservoir of electronic charge that, under appropriate circumstances depending on molecular geometry, is called on to stabilize bonds other than CH bonds by injecting electronic charge into the carbon skeleton, with a net gain in thermochemical stability.* This conclusion holds whenever the sum $\sum_l a_{kl}$, measuring the stabilization of all the bonds formed by atom k by an electronic charge added to it, is more negative than a_{HC}.

Cyclohexane is the example *par excellence*. For its boat form one calculates δ_C 10.7 for carbons 1 and 4 and δ_C 16.5 (ppm from ethane) for the other four

TABLE 13.2. Standard Enthalpy of Formation of Cycloalkanes (kcal/mol)

Molecule[a]		ΔE_a^*	ZPE + ΔH	ΔH_f° Calculated	ΔH_f° Experimental[b]
1	Cyclohexane	1760.85	107.72	−29.37	−29.50 ± 0.15
2	Methylcyclohexane	2056.75	125.64	−36.72	−36.98 ± 0.25
3	1,1-Dimethylch.	2351.99	143.57	−43.40	−43.23 ± 0.47
4	cis-1,2-Dimethylch.	2350.19	143.57	−41.06	−41.13 ± 0.44
5	trans-1,2-Dimethylch.	2352.04	143.57	−43.69	−42.99 ± 0.45
6	cis-1,3-Dimethylch.	2352.63	143.57	−44.05	−44.13 ± 0.42
7	trans-1,3-Dimethylch.	2350.61	143.57	−42.03	−42.18 ± 0.41
8	cis-1,4-Dimethylch.	2350.71	143.57	−42.13	−42.20 ± 0.42
9	trans-1,4-Dimethylch.	2352.62	143.57	−44.04	−44.10 ± 0.42
10	1-trans-2-cis-4-Me₃ch.	2645.73	161.50	−47.32	(−50)
11	1-cis-3-cis-5-Me₃ch.	2648.69	161.50	−51.55	−51.48
12	1-cis-3-trans-5-Me₃ch.	2646.75	161.50	−49.61	−49.37
13	Ethylcyclohexane	2350.26	143.80	−41.44	−41.05 ± 0.37
14	Propylcyclohexane	2644.08	161.96	−46.48	−46.20 ± 0.30
15	Isopropylcyclohexane	2645.04	161.64	−47.76	(−47)
16	n-Butylcyclohexane	2937.50	180.12	−51.11	−50.95 ± 0.33
17	1-Me-4-isopropylch.	2941.14	179.57	−55.30	−55.12 ± 0.80
18	n-Pentylcyclohexane	3231.09	198.28	−55.92	−55.88 ± 0.40
19	n-Hexylcyclohexane	3524.75	216.44	−60.79	−60.80 ± 0.43
20	n-Heptylcyclohexane	3818.43	234.60	−65.69	−65.73 ± 0.46
21	n-Octylcyclohexane	4112.14	252.76	−70.61	−70.65 ± 0.49
22	n-Nonylcyclohexane	4405.85	270.92	−75.54	−75.58 ± 0.54
23	n-Undecylcyclohexane	4993.28	307.24	−85.40	−85.43 ± 0.63
24	n-Tridecylcyclohexane	5580.72	343.56	−95.26	−95.28 ± 0.73
25	n-Pentadecylch.	6168.16	379.88	−105.12	−105.14 ± 0.83
26	Bicyclo[2.2.2]octane	2218.74	130.83	−24.09	−23.75 ± 0.30
27	trans-Decalin	2815.56	166.74	−43.77	−43.52 ± 0.56
28	cis-Decalin	2812.11	166.74	−40.32	−40.43 ± 0.56
29	Spiro[5.5]undecane	3103.70	183.20	−44.82	−44.81 ± 0.75
30	1,1′-Bicyclohexyl	3401.42	202.94	−52.18	−52.19 ± 0.74
31	trans−anti−trans-PHA	3865.08	225.76	−52.98	−52.74 ± 0.98
32	trans−syn−trans-PHA	3870.15	225.76	−58.05	−58.12 ± 0.93
33	Twistane	2679.14	153.95	−21.36	−21.6 ± 0.4
34	Adamantane	2688.15	153.95	−30.34	−30.65 ± 0.98
35	Diamantane	3615.93	199.50	−32.49	−32.60 ± 0.58

[a]The abbreviation "ch." stands for cyclohexane; PHA denotes perhydroanthracene.
[b]The experimental values of **1–9, 17, 27, 28,** and **32** are from Ref. 248; **34** is from Ref. 189, and **11** and **12** are from Ref. 262. The results for **13, 14,** and **18–25** are in Ref. 263, **30** is from Ref. 264, and **33** is from Ref. 265. The entries in parentheses are estimated values [266]. The ZPE + ΔH energies are calculated as indicated in Section 9.1. Calculated charges [36] were used for **15–25** as the ¹³C spectra were unavailable.

C atoms, using Grant's parameters [261]. Hence $\sum_k \delta_{C_k} = 87.4$, $\sum_k N_{C_k C} \delta_{C_k} = 174.8$ and [from Eq. (13.1)] $\Delta E_a^*(\text{boat}) = 1755.55 \text{ kcal/mol}$. For the chair form, the experimental value is δ_C 21.8 ppm from ethane, for $\sum_k \delta_{C_k} = 130.8$, $\sum_k N_{C_k C} \delta_{C_k} = 261.6$ and $\Delta E_a^*(\text{chair}) = 1760.86 \text{ kcal/mol}$, which is 5.31 kcal/mol better than $\Delta E_a^*(\text{boat})$. This result agrees with the measured ΔE_a^* energy increment (5.39 kcal/mol) between the *trans–anti–trans-* and *trans–syn–trans-* perhydroanthracenes, which differ only because of the center boat in the former compound, and with the difference, 5.23 kcal/mol, calculated from their ^{13}C spectra by means of Eq. (13.1).

A related example is offered by the *cis* and *trans* forms of 1,4-di-*t*-butyl-cyclohexane. Using the ^{13}C data of Roberts et al. [164] reported in Ref. 138, one finds $\sum_k \delta_{C_k} = 359.8$ and $\sum_k N_{C_k C} \delta_{C_k} = 778.6$ for the *trans* form. For the *cis-* form, on the other hand, it is $\sum_k \delta_{C_k} = 333.4$ and $\sum_k N_{C_k C} \delta_{C_k} = 716.8$. Hence $\Delta E_a^*(trans) = 4117.24 \text{ kcal/mol}$ for *trans*-1,4-di-*t*-butylcyclohexane and $\Delta E_a^*(cis) = 4113.72 \text{ kcal/mol}$ for the *cis* form. These two results are indicative of ring conformation since *cis*-1,4-di-*t*-butylcyclohexane is undoubtedly in a twist–boat form while the other is in chair conformation. The ^{13}C spectra of *t*-butylcyclohexane (in chair conformation) and of *trans*-1,4-di-*t*-butylcyclohexane are indeed very similar, except, of course, for carbon 4, which is the same as carbon 1 in the disubstituted molecule, whereas it is similar to the unsubstituted carbons in the monosubstituted cyclohexane.

It is by now clear by now that Eq. (13.1) performs well for all sorts of six-membered saturated carbon rings.

Smaller cycles, however, are definitely not described by this formula; the charge—NMR shift relationship (6.8) no longer applies and the a_{kl} parameters would also require modifications in order to adapt to the new situation. Of course, this does not come as a surprise.

What seems intriguing under the circumstances is a detail in the interpretation of ring strain. Present results (Table 1.1) suggest that the CC bond energy in cyclopropane is practically that of ethane—which seems odd, at first sight. Note that if we construct cyclopropane with the CC and CH bond energies of ethane, 69.63 and 106.81 kcal/mol, respectively, we get an estimated ΔE_a^* of 849.75 kcal/mol, which is alsmost the experimental value: 851.0 kcal/mol.

Now, where has the ring strain gone?

The point is that ring strain is defined by reference to cyclohexane. Its CC and CH bond energies are higher than those of ethane, by 3.15 and 3.47 kcal/mol, respectively. (This is due to electronic charge rearrangements ensuring electroneutrality of each CH_2 group.) As a consequence, cyclohexane is more stable by as much as 60.6 kcal/mol than what one would predict by making a simple sum of ethane bond energies, $6 \times 69.63 + 12 \times 106.81 \text{ kcal/mol}$. If the same were true for cyclopropane, it would be $\sim 30.3 \text{ kcal/mol}$ more stable than $3 \times 69.63 + 6 \times 106.81 \text{ kcal/mol}$, but this is precisely what does not happen. The bond energies remain roughly as they are in ethane—hence the ring strain, with respect to cyclohexane, of 29.4 kcal/mol.

This being said, it is a fact that Eq. (13.1) permits highly accurate calculations of molecular atomization energies, eventually leading to accurate evaluations of standard enthalpies of formation, ΔH_f° (298.15, gas). The point is that this approach requires only ^{13}C NMR shift results as substitutes for carbon atomic charges. This means that accurate rules for predicting these shifts, like those of Grant et al., for example, can be used as fast tracks to get access to reliable estimates of ΔH_f° energies. This potential could serve as a pretext for, and renew interest in, future extensive computer-assisted overhaulings and fine-tunings of all extant rules and proposals aimed at predicting accurate ^{13}C NMR shifts.

CHAPTER 14

UNSATURATED HYDROCARBONS

14.1 OLEFINS

The following presentation uses Eq. (10.31)

$$\sum_{k<l} \varepsilon_{kl} = \sum_{k<l} \varepsilon_{kl}^{\circ} + \sum_{k} \sum_{l} a_{kl}\Delta q_k + F$$

and the approximation $\Delta E_a^* \simeq \sum_{k<l} \varepsilon_{kl}$. The evaluation of nonbonded terms is not contemplated because the straightforward use of Eq. (10.3) cannot be justified for unsaturated hydrocarbons.

Reference Bond Energies

The $\sum_{k<l}\varepsilon_{kl}^{\circ}$ part is constructed from $\varepsilon_{CC}^{\circ} = 69.633$, $\varepsilon_{CH}^{\circ} = 106.806$ and $\varepsilon_{C=C}^{\circ} = 139.37$ kcal/mol, which are the references defined in Table 11.2 for ethane and ethylene, respectively.

Reference Charges

Both $q_C^{C_2H_6} = 35.1$ me and $q_C^{C_2H_4} = 7.7$ me are the selected net carbon charges of ethane and ethylene, respectively. The difference between them is

$$\Delta q_C^{\circ} = q_C^{C_2H_4} - q_C^{C_2H_6} = -27.4 \text{ me} \tag{14.1}$$

Atomic Charges, Bond Properties, and Molecular Energies, by Sándor Fliszár
Copyright © 2009 John Wiley & Sons, Inc.

The sp^3 carbons are simply written C. The sp^2 carbons are identified as $C(sp^2)$. Hence

$$\Delta q_C = q_C - q_C^{C_2H_6} \qquad \text{for } sp^3 \text{ C atoms} \tag{14.2}$$

$$\Delta q_{C(sp^2)} = q_{C(sp^2)} - q_C^{C_2H_4} \qquad \text{for } sp^2 \text{ C atoms} \tag{14.3}$$

Finally, the difference

$$q_{C(sp^2)} - q_C^{C_2H_6} = (q_{C(sp^2)} - q_C^{C_2H_4}) + (q_C^{C_2H_4} - q_C^{C_2H_6})$$

shows that

$$q_{C(sp^2)} - q_C^{C_2H_6} = \Delta q_{C(sp^2)} + \Delta q_C^\circ \tag{14.4}$$

where Δq_C° is the charge difference defined by Eq. (14.1).

For the hydrogen atoms, of course, $\Delta q_H = q_H - q_H^\circ$, where $q_H^\circ = -11.7$ me is the hydrogen net charge of ethane, selected as reference for *all* the H atoms. Their number is $2n$, with $n =$ number of the C atoms in the molecule. Charge normalization, $\sum q_H = -\sum q_C - \sum q_{C(sp^2)}$, and Eqs. (14.1)–(14.4) give

$$\sum \Delta q_H = -\left(\sum \Delta q_C + \sum \Delta q_{C(sp^2)}\right) - 2\Delta q_C^\circ + n q_H^\circ \tag{14.5}$$

Equation (14.5) conveniently eliminates explicit calculations of hydrogen charges from the general expression for $\sum_k \sum_l a_{kl} \Delta q_k + F$.

General Formula for Olefins

In the following, N_{CC} is the number of CC bonds and N_{CH} is the number of CH bonds formed by an sp^3 carbon atom. $N_{C(sp^2)C}$ and $N_{C(sp^2)H}$ are the numbers of CC and CH bonds, respectively, formed by an sp^2 carbon atom. Also note that $N_{CC} + N_{CH} = 4$ and that $N_{C(sp^2)C} + N_{C(sp^2)H} = 2$. Now we use Eq. (10.31). For the carbons we have the following equation, with $C' = C(sp^2)$ and remembering that $\Delta q_{C'} = \Delta q_{C'}^\sigma + \Delta q_{C'}^\pi$ (Chapter 11.2):

$$\sum_k^{CC} \sum_l^{bonds} a_{kl} \Delta q_k = a_{CC} \sum N_{CC} \Delta q_C + a_{C=C} \sum \Delta q_{C'}$$
$$+ a_{C'C} \sum N_{C'C}(\Delta q_{C'} + \Delta q_C^\circ) \tag{14.6}$$

Similarly, we obtain for the CH bonds that

$$\sum_k^{CH} \sum_l^{bonds} a_{kl} \Delta q_k = a_{CH} \sum N_{CH} \Delta q_C + a_{HC} \sum \Delta q_H$$
$$+ a_{C'H} \sum N_{C'H}(\Delta q_{C'} + \Delta q_C^\circ) \tag{14.7}$$

For F, we use F_{CC} (described as F in Example 11.1) and F_{CH} (Example 11.2) and obtain the simple result

$$F = F_{CC} \sum N_{C'C} + F_{CH} \sum N_{C'H} \tag{14.8}$$

The final energy formula follows from Eqs. (11.14)–(11.16) and (14.3)–(14.8):

$$\sum_k \sum_l a_{kl} \Delta q_k + F = A_1 \sum N_{C_kC} \Delta q_C + A_2 \sum \Delta q_{C_k} + a_{HC} n q_H^\circ$$

$$+ A_{1(C')}^{\sigma\pi} \sum N_{C'C} \Delta q_{C'} + A_3^{\sigma\pi} \sum \Delta q_{C'}$$

$$+ \left[\left(a_{C'C}^\sigma - a_{C'H}^\sigma \right) \Delta q_C^\circ + F_{CC} - F_{CH} \right] \sum N_{C'C}$$

$$+ \left[4F_{CH} + \left(4a_{C'H}^\sigma - 2a_{HC} \right) \Delta q_C^\circ \right] \tag{14.9}$$

where

$$A_1 = a_{CC} - a_{CH}$$
$$A_2 = 4a_{CH} - a_{HC}$$
$$A_{1(C')}^{\sigma\pi} = a_{C'C}^{\sigma\pi} - a_{C'H}^{\sigma\pi}$$
$$A_3^{\sigma\pi} = a_{C=C}^{\sigma\pi} + 2a_{C'H}^{\sigma\pi} - a_{HC}$$

Equation (14.9) lends itself to numerical tests (see Table 14.1). The first two terms are well known; they are those described for the saturated carbons (see Table 10.4); hence $A_1 = 0.0356$ and $A_2 = 0.0529$ kcal mol^{-1} ppm^{-1}. We also know that $a_{HC} nq_H^\circ = 7.393 n$ kcal/mol. Using this theoretical input and the appropriate sums, $\sum_{k<l} \varepsilon_{kl} \simeq \Delta E_a^*$, in comparisons with experimental atomization energies, one obtains $A_{1(C')}^{\sigma\pi}$ and $A_3^{\sigma\pi}$ and the empirical estimates of the two terms in brackets given in Table 14.1. The latter can be evaluated theoretically using the F_{CC} and F_{CH} results given in Examples 11.1 and 11.2, respectively, and the appropriate $a_{C'C}^\sigma(1.531)$ and $a_{C'H}^\sigma(1.08)$ parameters described in Chapter 11 (Examples 11.5 and 11.11, respectively), with $\Delta q_C^\circ = -27.4$ me [Eq. (14.1)]. Incidentally, note that $a_{C'C}^\sigma - a_{C'H}^\sigma = a_{CC} - a_{CH} = -0.241$ kcal mol^{-1} me^{-1}. The empirical results nicely support the theoretical predictions. Also note that the latter cover the largest part, by far, of all the energy terms occurring in ΔE_a^*. It appears, indeed, that the unresolved part of Eq. (14.9), $A_{1(C')}^{\sigma\pi} \sum N_{C'C} \Delta q_{C'} + A_3^{\sigma\pi} \sum \Delta q_{C'}$, represents less than 1 kcal/mol.

TABLE 14.1. Tests of Eq. (14.9) for *trans*- and *cis*-Olefins (kcal/mol)

	trans-Olefins		*cis*-Olefins	
Parameter	Theoretical	Empirical	Theoretical	Empirical
$4F_{CH} + (4a_{C'H}^\sigma - 2a_{HC})\Delta q_C^\circ$	−19.08	−19.16	−18.27	−18.22
$(a_{C'C}^\sigma - a_{C'H}^\sigma)\Delta q_C^\circ + F_{CC} - F_{CH}$	4.17	4.19	3.97	4.0

The empirical evaluation of this small residual quantity indicates that $A^{\sigma\pi}_{1(C')}\Delta q_{C'} \simeq -0.028\delta_{C'}$ and $A^{\sigma\pi}_3\Delta q_{C'} \simeq 0.20\delta_{C'}$ kcal/mol. Knowing that $A^{\sigma\pi}_{1(C')} = -0.179$ and $A^{\sigma\pi}_3 = 1.357$ kcal mol^{-1} me^{-1} (from Table 11.3), one finds $m \simeq -0.955$, $d\delta_{C'}/dq^{\pi} \simeq 288$ ppm/e and $\Delta q_{C'} \simeq 0.15\delta_{C'}$ me [109].

Because of its little weight in the final results, we may as well simplify things and replace $A^{\sigma\pi}_{1(C')}\sum N_{C'C}\Delta q_{C'} + A^{\sigma\pi}_3\sum\Delta q_{C'}$ by $0.18\sum\delta_{C'}$ to cover that part of Eq. (14.9) [109]. The approximations (in kcal/mol units)

$$\sum_k\sum_l a_{kl}\Delta q_k \simeq 0.0356\sum N_{CC}\delta_C + 0.0529\sum\delta_C + 0.18\sum\delta_{C'}$$

$$+ 7.393n + 4.19\sum N_{C'C} - 19.16 \tag{14.10}$$

for ethylene, 1-alkenes, *trans*-alkenes, and tetramethylethylene, and

$$\sum_k\sum_l a_{kl}\Delta q_k \simeq 0.0356\sum N_{CC}\delta_C + 0.0529\sum\delta_C + 0.18\sum\delta_{C'}$$

$$+ 7.393\,n + 4.0\sum N_{C'C} - 18.22 \tag{14.11}$$

for *cis*-olefins appear to be quite adequate in practical applications (see Table 14.2). The idea of using approximate *transferable* F_{CC} and F_{CH} bond terms for general use is justified by the present results.

Dienes

The underlying idea is simple. The sum ε°_{kl} is calculated (see Table 11.2) using ε°_1 and ε°_{10} for the $C(sp^3)$—$C(sp^3)$ and $C(sp^3)$—H bonds, respectively, and ε°_2 for the double bonds. For the $C(sp^2)$—$C(sp^2)$ single bond, one takes ε°_5 and, finally, uses ε°_4 and ε°_{11} for the $C(sp^3)$—$C(sp^2)$ and $C(sp^2)$—H bonds, respectively. No separate calculation of F is required, for it is included in the modified references ε°_4, ε°_5, and ε°_{11}.

The calculation of $\sum_k\sum_l a_{kl}\Delta q_k$ involves the following steps:

1. First, we form the sum $\sum_l a_{kl}$ for each individual carbon atom and calculate the corresponding Δq_C. So we obtain $\Delta q_C \times \sum_l a_{kl} = \sum_l a_{kl}\Delta q_k$ for each C_k atom of the molecule. The sum of all these $\sum_l a_{kl}\Delta q_k$ terms gives the total contribution of all the C atoms to the final sum $\sum_k\sum_l a_{kl}\Delta q_k$.

2. Using the same Δq_C data, one forms their sum $\sum\Delta q_C$. Knowing the reference charges, 35.1 me for the sp^3 carbons and 7.7 me for the sp^2 C atoms, one finds the lump total net charge $\sum q_C$ of all the carbons.

3. Thus we have $\sum q_H = -\sum q_C$ and $\sum\Delta q_H = \sum q_H - n_H q^\circ_H$, where n_H is the number of hydrogen atoms. So we get $a_{HC}\sum\Delta q_H$, which is the total contribution of all the hydrogen atoms to $\sum_k\sum_l a_{kl}\Delta q_k$.

TABLE 14.2. Standard Enthalpy of Formation of Olefins (kcal/mol)

Molecule[a]		$\sum_k \sum_l a_{kl} \Delta q_k$	ΔE_a^*	ΔH_f° Calculated	ΔH_f° Experimental
1	Ethene	−4.37	562.22	12.38	12.50
2	Propene	8.80	858.64	4.80	4.88
3	1-Butene	19.47	1152.55	−0.27	−0.03
4	(Z)2-Butene	20.74	1153.83	−1.54	−1.67
5	(E)2-Butene	21.91	1154.99	−2.71	−2.67
6	2-Me-Propene	23.42	1156.51	−4.22	−4.04
7	1-Pentene	30.08	1446.41	−5.28	−5.00
8	(Z)2-Pentene	31.55	1447.88	−6.75	−6.71
9	(E)2-Pentene	32.39	1448.72	−7.60	−7.59
10	2-Me-1-Butene	33.84	1450.17	−9.05	−8.68
11	3-Me-1-Butene	31.33	1447.66	−6.87	−6.92
12	2-Me-2-Butene	34.98	1451.31	−10.19	−10.17
13	1-Hexene	40.18	1739.76	−9.80	−9.96
14	(Z)2-Hexene	42.02	1741.60	−11.64	−12.51
15	(E)2-Hexene	42.99	1742.57	−12.61	−12.88
16	(Z)3-Hexene	42.28	1741.86	−11.90	−11.38
17	(E)3-Hexene	43.00	1742.58	−12.62	−13.01
18	2-Me-1-Pentene	44.25	1743.82	−13.86	−14.19
19	3-Me-1-Pentene	41.65	1741.22	−11.60	−11.82
20	4-Me-1-Pentene	41.94	1741.52	−11.90	−12.24
21	(Z)3-Me-2-Pentene	45.71	1745.29	−15.33	−15.08
22	(E)3-Me-2-Pentene	45.14	1744.71	−14.75	−14.86
23	(Z)4-Me-2-Pentene	43.59	1743.16	−13.54	−13.73
24	(E)4-Me-2-Pentene	44.43	1744.00	−14.38	−14.69
25	2-Et-1-Butene	43.34	1742.91	−12.95	−13.38
26	2,3-diMe-1-Butene	45.37	1744.95	−15.33	−15.85
27	3,3-diMe-1-Butene	43.31	1742.89	−13.61	−14.70
28	2,3-diMe-2-Butene	47.17	1746.74	−16.78	−16.68
29	(Z)2-Heptene	52.47	2035.29	−16.49	−16.9
30	(E)2-Heptene	53.43	2036.25	−17.44	−17.6
31	(Z)3-Heptene	52.76	2035.58	−16.77	−16.90
32	(E)3-Heptene	53.59	2036.40	−17.60	−17.60
33	(Z)3-Me-3-Hexene	55.85	2038.66	−19.86	−18.60
34	(E)3-Me-3-Hexene	55.92	2038.74	−19.93	−19.22
35	2,4-diMe-1-Pentene	56.38	2039.19	−20.74	−20.27
36	4,4-diMe-1-Pentene	54.65	2037.47	−19.55	−19.20
37	(E)4,4-diMe-2-Pentene	56.98	2039.70	−21.58	−21.46
38	(E)2,2-diMe-3-Hexene	67.45	2333.52	−26.56	−26.16
39	2-Me-3-Et-1-Pentene	66.06	2332.12	−24.82	−24.40

[a]Note that the $\sum_k \sum_l a_{kl} \Delta q_k$ results *include* F. The experimental ΔH_f° values are taken from Ref. 242. The NMR data are from Refs. 138, 267, and 268.

TABLE 14.3. Comparison between Calculated and Experimental Atomization Energies of Selected Dienes (kcal/mol)

Molecule[a]	ΔH_f°	$\sum_k \sum_l a_{kl} \Delta q_k$	ΔE_a^* Calculated	ΔE_a^* Experimental
1,3-Butadiene	26.33	−21.57	1010.4	1010.1
(Z)1,3-Pentadiene	19.77	−14.09	1305.3	1305.4
(E)1,3-Pentadiene	18.77	−12.85	1306.6	1306.5
Isoprene	18.06	−13.09	1306.3	1306.7
1,4-Pentadiene	25.25	−11.61	1300.2	1300.1
1,5-Hexadiene	20.11	−0.99	1594.1	1594.0
Dimethyl-1,3-butadiene	10.78	−4.20	1602.6	1602.8
1,3-Cyclohexadiene	25.38	10.98	1473.8	1473.7

[a]The sources of the experimental ΔH_f° values and of the NMR shifts used in these calculations are reported in Ref. 192.

4. The final sum $\sum_k \sum_l a_{kl} \Delta q_k$ is obtained by adding the contributions of the carbon and the hydrogen atoms.

This approach is probably the most convenient one for general use. It could have been applied to the saturated and ethylenic hydrocarbons examined earlier; as a rule, one can always use the atom-by-atom method described here, with modifications, if necessary, depending on the class of molecules investigated.

The final results are collected in Table 14.3.

14.2 AROMATIC MOLECULES

The calculations are best made following the strategy explained for the dienes, in the atom-by-atom mode.

The $\sum \varepsilon_{kl}^\circ$ part is constructed using the reference bond energies of Table 11.2. A comment is in order regarding ε_8°, for a bond between aryl carbons at a distance of 1.397 Å. Consider the aromatic CC bond, ε_3°. It represents, so to speak, the "average" between a single bond and a double bond as they are found in benzenoid structures; it is the CC bond of benzene. It is counted twice the number of double bonds one can write using classical Kekulé structures, for instance, 10 times for naphthalene and 14 times for anthracene. The remaining CC bonds (e.g., one in naphthalene, two in anthracene) are treated as $C(sp^2)$—$C(sp^2)$ single bonds; these are the bonds described by ε_8°. Of course, no bond in particular is identified in this manner; it is only the number of bonds that matters. It is now clear that one cannot simply use ε_3° for each CC bond found in aromatic cycles because there are not as many "averages" as there are CC bonds, except in benzene itself.

All this is best illustrated by an example: 2-methylnaphthalene. The presence of the CH_3 group offers the opportunity of using Eq. (11.17). The charges of the carbon atoms are deduced from their NMR shifts, using Eq. (6.6) for the aromatic carbons and (6.9) for the CH_3 carbon.

Example 14.1: Atom-by-Atom Calculation of 2-Methylnaphthalene. Using the Δq_C values calculated from the ^{13}C shifts, counting 13.2 me for each aromatic carbon and 35.1 me for the sp^3 reference, we get $\Sigma q_C = 162.88$ me and $\Sigma \Delta q_H = -162.88 - 10 \times (-11.7) = -45.88$ me. Note the contribution of carbon-2, $-0.060\delta_C = -0.408$ kcal/mol. The sum over all the atoms k of all their $\Sigma_l a_{kl}\Delta q_k$ terms yields 30.00 kcal/mol, ready for use in Eq. (10.37). The sum $\sum_{k<l} \varepsilon^{\circ}_{kl}$ is, using the entries of Table 11.2, $10\varepsilon^{\circ}_3 + \varepsilon^{\circ}_8 + \varepsilon^{\circ}_6 + 7\varepsilon^{\circ}_{12} + 3\varepsilon^{\circ}_{10} = 2424.73$ kcal/mol, so that ΔE^*_a 2454.7_3 kcal/mol. (See Table 14.4.)

The final results are reported in Table 14.5.

In the absence of experimental results, predicted ΔH°_f values are indicated in parentheses. Two ΔE^*_a results listed in the column reporting "experimental values" are indicated in parentheses; these are theoretical results offered for comparison, deduced from enthalpies of formation calculated by Dewar and de Llano [269].

The unsigned average deviation between calculated and experimental energies is 0.36 kcal/mol for a collection of 35 benzenoid molecules. This result does not include 7,12-dimethylbenz[a]anthracene (**36**): the discrepancy of ~16 kcal/mol between theory and experiment is in all likelyhood due in part to an error in the latter. Although certainly real, steric interactions involving the methyl group in position 12 are probably not so severe as to cause a destabilization exceeding that found in 1,8-dimethylnaphthalene and 4,5-dimethylphenanthrene—molecules that are discussed further below.

TABLE 14.4. Atom-by-Atom Evaluation of $\sum_{k,l} a_{kl} \Delta q_k$ for 2-Methylnaphthalene (kcal/mol)[a]

Atom k	δ_C	Δq (me)	$\Sigma_l a_{kl}$	$\Delta q_k \Sigma_l a_{kl}$
1	-1.5	-1.80	-0.799	1.438
2	6.8	—	(-0.060)	-0.408
3	-0.45	-0.54	-0.799	0.431
4	-0.7	-0.84	-0.799	0.671
5	-0.7	-0.84	-0.799	0.671
6	-3.6	-4.32	-0.799	3.452
7	-2.7	-3.24	-0.799	2.589
8	-1.1	-1.32	-0.799	1.055
9	5.6	6.72	-1.074	-7.217
10	3.65	4.38	-1.074	-4.704
CH_3	16.35	-2.42	-1.229	2.974
$\sum \Delta q_H$	—	-45.88	-0.632	28.996

[a]The δ_C data are relative to benzene (1–10) and to ethane for the methyl carbon.

TABLE 14.5. Calculated and Experimental ΔE_a^* Energies of Benzenoid Hydrocarbons (kcal/mol)

				ΔE_a^*	
Molecule		ΔH_f°	ZPE + ΔH	Calculated	Experimental
1	Benzene	19.81 ± 0.13	66.22	1366.5	1366.5
2	Toluene	11.99 ± 0.10	84.43	1663.4	1663.2
3	1,2-diMe-BZ	4.56 ± 0.26	102.64	1959.8	1959.4
4	1,3-diMe-BZ	4.14 ± 0.18	102.64	1960.1	1959.9
5	1,4-diMe-BZ	4.31 ± 0.24	102.64	1959.3	1959.7
6	1,2,3-triMe-BZ	−2.26 ± 0.29	120.85	2256.4	2255.1
7	1,2,4-triMe-BZ	−3.31 ± 0.26	120.85	2255.7	2256.2
8	1,3,5-triMe-BZ	−3.81 ± 0.33	120.85	2256.6	2256.6
9	1,2,3,4-tetraMe-BZ	−10.02	139.06	2552.5	2551.7
10	1,2,3,5-tetraMe-BZ	−10.71	139.06	2552.1	2552.4
11	1,2,4,5-tetraMe-BZ	−10.81	139.06	2551.4	2552.5
12	Pentamethyl-BZ	−17.80	157.27	2847.9	2848.3
13	Hexamethyl-BZ	−25.26	175.48	3144.6	3144.6
14	Ethylbenzene	7.15 ± 0.19	102.64	1956.7	1956.9
15	n-Propylbenzene	1.89 ± 0.19	120.85	2250.3	2250.9
16	Isopropylbenzene	0.96 ± 0.26	120.50	2251.5	2251.5
17	sec-Butylbenzene	−4.15 ± 0.31	138.71	2545.0	2545.5
18	tert-Butylbenzene	−5.40 ± 0.31	138.71	2547.4	2546.7
19	1,2-Diphenylethane	32.4 ± 0.3	155.0	3203.0	3202.9
20	Styrene	35.30 ± 0.25	85.80	1811.5	1810.7
21	cis-Stilbene	60.31 ± 0.42	138.05	3056.8	3056.8
22	trans-Stilbene	52.5	138.05	3064.0	3064.6
23	Biphenyl	43.53 ± 0.60	118.27	2613.2	2613.7
24	Naphthalene	36.25 ± 0.45	94.90	2157.7	2157.6
25	1-Me-naphthalene	27.93 ± 0.64	113.11	2455.3	2454.8
26	2-Me-naphthalene	27.75 ± 0.62	113.11	2454.7	2454.9
27	1,8-diMe-naphthalene	See text	131.32	2745.0	2745.0
28	Anthracene	55.2 ± 1.1	123.7	2946.4	2946.3
29	9-Methylanthracene	(42.1)	141.8	3248.1	—
30	9,10-diMe-anthracene	(31.6)	160.0	3547.5	—
31	Phenanthrene	49.5 ± 1.1	123.7	2952.7	2952.0
32	Pyrene	53.94 ± 0.31	133.05	3295.7	3295.7
33	Triphenylene	61.9 ± 1.1	152.3	3746.5	3747.1
34	Benz[a]anthracene	65.97	152.3	3743.0	3743.0
35	7-Me-benz[a]AN	(56.2)	170.5	4041.6	—
36	7,12-diMe-benz[a]AN	66.4 ± 1.1	188.7	4336.0	4320.2
37	Dibenz[a,c]AN	(77.2)	181.0	4539.3	(4540.5)
38	Dibenz[a,h]AN	(79.6)	181.0	4536.9	(4538.0)
39	1,2,3,4-TetrahydroNA	7.3 ± 1.3	126.1	2420.2	2420.2
40	9,10-DihydroAN	38.2 ± 1.1	142.3	3082.8	3083.1
41	9,10-DihydroPHE	(37.8)	142.3	3083.5	—

(Continued)

TABLE 14.5. *Continued*

Molecule		ΔH_f°	$ZPE + \Delta H$	ΔE_a^* Calculated	ΔE_a^* Experimental
42	(1–8)-OctahydroAN	(−4.3)	185.9	3472.9	—
43	(1–8)-OctahydroPHE	(−4.9)	185.9	3473.5	—
44	(1–12)-DodecaHTPH	(−16.4)	245.8	4526.2	—

Key: BZ = benzene (**3–13**); AN = anthracene (**35–38, 40, 42**); NA = naphthalene (**39**); PHE = phenanthrene (**41, 43**); HTPH = hydrotriphenylene (**44**).
Sources: The sources of the experimental ΔH_f° and NMR data are indicated in Ref. 129.

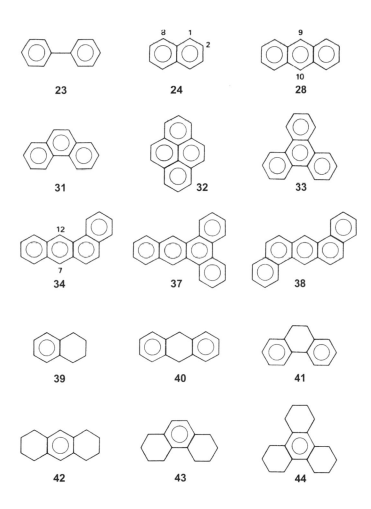

The general ideas of our energy analysis are certainly supported by the results. On the other hand, they should not hide the limitations of our approach due to geometry-related problems. They are best revealed by the following examples.

Example 14.2: The *trans*- and *cis*-Stilbenes. *trans*-Stilbene was treated as a planar system. The $C(sp^2)$—Ph bonds were accordingly derived on the basis of a conjugated sp^2–sp^2 single bond. For styrene (Example 11.8), we found $\varepsilon_7^\circ = 89.69$ for its $C(Ar)$—$C(sp^2)$ bond, with $F = -0.78$. For *trans*-stilbene (with $R_{CC} = 1.48$ Å and $\varphi = 125°$), the same calculation gives $F = -0.28$. Moreover, the charge variation from 13.2 to 7.7 me required $a_{CC}^\sigma(1.445) = -0.472$ in styrene but $a_{CC}^\sigma(1.48) = -0.463$ in *trans*-stilbene. Hence $\varepsilon_{7'}^\sigma = \varepsilon_7^\circ + 0.78 - 0.28 - 0.05 = 90.14$ kcal/mol.

Now, while the molecular structure of *trans*-stilbene, C_6H_5—CH=CH—C_6H_5, apprears to be approximately planar in the solid state [270], its gas-phase structure is found to be nonplanar [271]. However, the potential curve for energy versus the dihedral angle is very shallow and the calculated energy barrier corresponding to the C_i conformation is only about 0.5 kcal/mol [233]. Both these results and our calculation suggest that in *trans*-stilbene there is no great energy difference between conjugation and hyperconjugative stabilization of the sp^2–sp^2 single bond, but it is also clear that in this particular situation it could not be assumed a priori that our approach would lead to a valid result, as it did.

The case of *cis*-stilbene is clear-cut. Electron diffraction data [272] point to a C_2 symmetry in the gas phase and a structure that may be described as having a propellerlike conformation with phenyl groups rotated $\sim$45° about the C—Ph bonds. The latter were thus treated as nonconjugated bonds, with $R_{CC} = 1.49$ Å [272], $\Delta r_{\pi l} = 0.029$ Å, and $\varphi = 129°$, for $F = -4.65$, by reference to the nonconjugated central CC bond of biphenyl. The charge variation from 13.2 to 7.7 me at the sp^2 carbon of the olefinic part, calculated with $a_{CC}^\sigma(1.49) = -0.460$, contributes 2.53 kcal/mol. Thus we deduce for the C—Ph bonds of *cis*-stilbene that $\varepsilon_{7''}^\sigma = 88.89 + 2.53 - 4.65 = 86.77$ kcal/mol. Similarly, the gas-phase value [273] for the tortional angle about the central bond of biphenyl, 41.6°, and its bond length, $R = 1.49$ Å [233], suggest that the central bond should be treated like a nonconjugated CC single bond, which led to $\varepsilon_9^\circ = 88.89$ kcal/mol^{-1}.

Example 14.3: Triphenylene. The same situation arises with triphenylene (**33**), which is significantly nonplanar [274]. The bonds joinning the "external" rings to one another were thus calculated at $\varepsilon_{CC} = 88.45$ kcal/mol for $R = 1.46$ Å, following the approach used for biphenyl. It is clear that some advance knowledge is necessary in our calculations, namely, regarding planarity (or lack of it) of the benzenoid skeleton.

Example 14.4: 4,5-Dimethylphenanthrene. Using the ^{13}C NMR spectrum measured by Stothers et al. [275], we deduce $\Delta H_f^\circ = 36.8$ kcal/mol assuming planarity. Closely neighboring methyl groups that are separated by five bonds in the molecular skeleton, however, result in chiral nonplanar conformations [276]. Modeling, where appropriate, the CC bonds on those described for biphenyl and *cis*-stilbene, one predicts $\Delta H_f^\circ = 47.8$ for the nonplanar form, in acceptable accord with the reported value [248], 46.26 ± 1.46 kcal/mol, a result that is self-explanatory.

Example 14.5: The Dimethylnaphthalene Isomers. For the isomer with the substituents in the 1,8 position, our calculation yields $\Delta E_a^* = 2748.6\,\text{kcal/mol}$ for the planar form, in error with respect to its experimental counterpart,[1] $2745.0\,\text{kcal/mol}$. The thermochemical stability is overestimated by $\sim 3.6\,\text{kcal/mol}$, thus suggesting a possible loss of conjugation in this molecule, which, in fact, is known to suffer considerable distortion from the normal naphthalene geometry [278]. Indeed, a calculation following the lines described above for biphenyl and non-planar 4,5-dimethylphenanthrene agrees with experiment [129]. In contrast, 2,6-dimethylnaphthalene can safely be assumed to retain the planar geometry of naphthalene. The result deduced for this form, $\Delta E_a^* = 2751.3$, agrees well with the experimental value, $2751.0\,\text{kcal/mol}$.

The examples illustrate possible applications of energy analyses based on ^{13}C spectra in problems regarding the origin of molecular stability, namely in terms of a partial suppression of conjugation accompanying deformations of a benzenoid skeleton. The great diversity of bonds formed by sp^3 and sp^2 carbons and conjugation have been dealt with efficiently. Chemical insight and accuracy reflect, and benefit from, the simplicity embodied in Eqs. (10.35) and (11.13).

Beyond all complications that seem to accompany the multitude of possible carbon–carbon bonds, simple familiar intuition is vindicated; it is not false, after all, to consider the C(Ar)—C(Ar) bonds of benzene as a sort of average between ε_2° (the double bond of ethylene) and a single CC bond, provided the latter is chosen properly: namely, the conjugated sp^2–sp^2 single bond, ε_8°, between aromatic carbons (in lieu of the CC single bond of ethane).

The proper use of the "average" aromatic bond ε_3° coupled with that of the supernumerary "aromatic single bonds." ε_8° is clearly illustrated in the following example.

Example 14.6: Graphite. Graphite has a layerlike structure. Each carbon is bonded to three other carbons forming a framework of planar benzenoid rings, with bond lengths of 1.42 Å [279]. Two of these bonds command the use of ε_3°, while ε_8° must be used for the third one. Now we calculate $a_{CC}\,\Delta q_C$. The graphite carbons are electroneutral; hence $\Delta q_C = -13.2$ me. Using $a_{CC}^{\sigma\pi}(1.42) = -0.352\,\text{kcal/mol}^{-1}$ me^{-1} (for $m = -0.814$), one finds

$$\Delta E_a^*\ (\text{graphite}) = \frac{1}{2}\left(2\varepsilon_3^\circ + \varepsilon_8^\circ + 3 \times 9.29\right) = 174.93\,\text{kcal/mol}$$

The energy of atomization at $T = 298.15$ K, namely $\Delta E_a = \Delta H_a^\circ - RT$ (where ΔH_a° is the corresponding enthalpy), thus follows from ΔE_a^* by subtracting the zero-point and the gas-phase heat content $(H_T - H_0)$ and by adding the translational energy $(\frac{3}{2}RT)$ of the carbon atoms formed during the atomization of graphite: $\Delta H_a^\circ = \Delta E_a^* - \text{ZPE} - (H_T - H_0) + \frac{5}{2}RT$. The standard enthalpy of atomization

$$\Delta H_a^\circ\ (\text{graphite, 298.15 K}) = 171.17\,\text{kcal/mol}$$

---

[1]These results follow from thermochemical data at 350–370 K given in Ref. 277.

is obtained with the help of the calculated ZPE and the recommended gas-phase heat content, 3.68 [199] and 1.562 ± 0.002 [280] kcal/mol, respectively. Our calculation has neglected the interactions between the layers, of the order of ~ 1.2 kcal/mol [281], but is nonetheless in acceptable agreement with the experimental value, $\Delta H_a^\circ = 171.29 \pm 0.11$ kcal/mol (gas, 298.15 K) [280].

This calculation of graphite represents a severe test of the theory; because of the large weight of $a_{CC}^{\sigma\pi} \Delta q_C$ in the final results, ~ 13.94 kcal/mol, it seems fair to conclude that the charge assigned to the benzene carbon, 13.2 me, should be correct at least within ± 1 me.

CHAPTER 15

NITROGEN-CONTAINING MOLECULES

15.1 AMINES: CHARGES OF THE CARBON ATOMS

For sp^3 carbon atoms, the relationship (6.8) with ^{13}C nuclear magnetic resonance shifts is highly accurate [38]

$$q_C = -0.148(\delta_C - \delta_C^{C_2H_6}) + 35.1 \,(\text{me})$$

where δ_C and $\delta_C^{C_2H_6}$, the shift of the ethane carbon, are relative to TMS. For the alkyl-amines, we can use the ^{13}C NMR shifts of Eggert and Djerassi [165]. This aspect has been tested carefully [139].

Alkane carbon atoms satisfy the charge–NMR shift correlation [Eq. (6.8)]. With the alkylamines, things could be different because of a possible "extra" effect due to the presence of the nitrogen atom: α-carbons should perhaps be compared only among themselves, and so should the β- and γ-carbons. The δ-carbons, in contrast, which are sufficiently separated from the nitrogen center, could probably be treated as if they were part of an alkane. This point has been examined as follows for the —$C_\beta H_2$—$C_\alpha H_2$—NH_2 motif, focusing on the dissociation and intrinsic bond energies, $D_{C_\alpha C_\beta}$ and $\varepsilon_{C_\alpha C_\beta}$, respectively.

The first step regards the reorganizational energy of the $NH_2 CH_2 \cdot$ radical—it is known (Table 12.3).

This reorganizational energy, the ΔE_a^* energies deduced from experimental enthalpies [248], and the energies of the alkyl radicals (Table 12.3) now give the following

Atomic Charges, Bond Properties, and Molecular Energies, by Sándor Fliszár
Copyright © 2009 John Wiley & Sons, Inc.

dissociation energies, $D_{C_\alpha C_\beta}$ of the $C_\alpha C_\beta$ bonds (in kcal/mol): 89.47 (ethylamine), 88.89 (propylamine) and 88.80 (butylamine) (Table 12.5). Thus the corresponding intrinsic energies, $\varepsilon_{C_\alpha C_\beta}$, Eq. (12.9): 71.99, 73.81, and 74.58 kcal/mol, respectively. A recalculation of the same $C_\alpha C_\beta$ bonds, using the ^{13}C chemical shifts of Ref. 165 in Eq. (6.8) to obtain the charges, gave 72.83, 73.99, and 74.51 kcal/mol, respectively. A similar result is also obtained for 1,2-diaminoethane (see Table 12.5).

It is difficult to assess how close the two sets of results really are. The first one evidently depends on the precision of the thermochemical data that have been used, namely, ΔH_f° and $ZPE + H_T - H_0$. Equation (6.8), on the other hand, is accurate. It is perhaps our best means for testing sp^3 carbon charges; an error of 1 me in the evaluation of one of the carbons translates into an error of ~0.5 kcal/mol in bond energy. Now, the two sets are too close to warrant revision of the procedure, yet we cannot endorse it for more than it is: an acceptable approximation. For our needs, and for the time being, Eq. (6.8) solves the problem. Moreover, the reasoning is that if this approximation holds in the close neighborhood of nitrogen, it should be all the more acceptable for carbons in positions γ, δ, and so on.

15.2 NITROGEN CHARGES AND BOND ENERGIES

The electrostatic potentials V_H at the nuclei of hydrogen atoms bonded to nitrogen and carbon atoms offer much required information regarding a few key molecules, namely, about the energies of their chemical bonds.

These potentials, for use in Eq. (10.15), are efficiently computed with the help of density functional (DFT) methods, namely, Becke's approach [282,184] with the Lee–Yang–Parr (LYP) potential [20]. Here we adopt a fully coherent method in which the SCF process, the geometry optimizations, and the computation of analytic second derivatives are carried out with the complete density functional, including gradient corrections and HF exchange. The performance of this three-parameter density functional, hereafter referred to as B3LYP, was investigated earlier, namely in thermochemistry [283,284]. We selected Pople's 6-311G** basis [16].

The thus computed potential energies, V_k (Eq. (10.14)), were rescaled with the help of the *experimental* total energy, using the Politzer formula [79]

$$E^{\text{molecule}} = \sum_k \frac{V_k}{\gamma_k^{\text{mol}}}$$

and the appropriate γ_k^{mol}s determined from B3LYP/6-311G** calculations, namely, $\gamma_H = 2$, $\gamma_C = 2.322864$ and $\gamma_N = 2.343435$ [207]. The rescaled values

$$V_k^{\text{rescaled}} = \frac{V_k^{\text{B3LYP}} \times E_{\text{exper}}^{\text{molecule}}}{\sum_k (V_k/\gamma_k^{\text{mol}})}$$

differ little from the original values (see Tables 15.1 and 15.3).

TABLE 15.1. Electrostatic Potential at H Nuclei

Molecule		Potential at H, V_H (au)	
		B3LYP/6-311G**	Rescaled
NH_3		-1.072288	-1.070521
NH_2-NH_2		-1.075520	-1.073657
CH_3-CH_3		-1.132693	-1.133453
CH_3-NH_2	at CH_3	-1.132237	-1.131434
	at NH_2	-1.077308	-1.076544
$(CH_3)_2NH$	at CH_3	-1.128969	-1.128862
	at NH	-1.077484	-1.077382
$(CH_3)_3N$		-1.127187	-1.127573

Selected computed data are collected in Table 15.1. They are of great use in the forthcoming examples, which illustrate the procedure.

Example 15.1. The NH bond energy in NH_3, one-third of its atomization energy $\Delta E_a^* = 297.31$ kcal/mol, is $\varepsilon_{NH} = 99.10$ kcal/mol, with $V_H = -1.070521$ au (Table 15.1). For the hydrogen atoms in NH_2-NH_2, we find $V_H = -1.073657$ au. Equation (10.15) thus suggests that its NH bond energy differs by $\Delta\varepsilon_{NH} = -\frac{1}{2}(-1.073657 + 1.070521)$ au $= 0.984$ kcal/mol from that of ammonia. Hence, in this approximation, $\varepsilon_{NH} = 100.08$ kcal/mol in hydrazine, and finally, considering its atomization energy $\Delta E_a^* = 436.62$ kcal/mol, we find that $\varepsilon_{NN} = 36.30$ kcal/mol.

Example 15.2. The CH bond energy in C_2H_6 is 106.806 kcal/mol with $V_H = -1.133453$ au. In CH_3NH_2, we find $V_H = -1.131434$ au (on average) for the CH_3 hydrogens and thus $\Delta\varepsilon_{CH} = -0.633$ kcal/mol and $\varepsilon_{CH} = 106.17$ kcal/mol in methylamine.

TABLE 15.2. Energy Parameters of Selected Molecules (kcal/mol)

Molecule	ΔE_a^*	Bond Energy
NH_3	297.31	$\varepsilon_{NH} = 99.10$
NH_2-NH_2	436.62	$\varepsilon_{NH} = 100.08$
		$\varepsilon_{NN} = 36.30$
CH_3-NH_2	580.95	$\varepsilon_{NH} = 100.99$
		$\varepsilon_{CH} = 106.17$
		$\varepsilon_{CN} = 60.44$
$(CH_3)_2NH$	868.70	$\varepsilon_{NH} = 101.25$
		$\varepsilon_{CH} = 105.37$
		$\varepsilon_{CN} = 67.63$
$(CH_3)_3N$	1158.66	$\varepsilon_{CH} = 104.96$
		$\varepsilon_{CN} = 71.34$

TABLE 15.3. Electrostatic Potential at Alkyl-H Nuclei (au) and CH Bond Energies (kcal/mol)

Molecule	Site	Potential[a] at H, V_H (au)		ε_{CH} Energy
		B3LYP/6-311G**	Rescaled	
$C_2H_5NH_2$	α-CH$_2$	-1.134632	-1.134240	107.05
	β-CH$_3$	-1.125653	-1.125264	104.24
$C_3H_7NH_2$	α-CH$_2$	-1.133008	-1.132833	106.61
	β-CH$_2$	-1.131283	-1.131108	106.07
	γ-CH$_3$	-1.128297	-1.128123	105.13
n-C$_4$H$_9$NH$_2$	α-CH$_2$	-1.135102	-1.134904	107.26
	β-CH$_2$	-1.132334	-1.132137	106.39
	γ-CH$_2$	-1.133984	-1.133786	106.91
	δ-CH$_3$	-1.129626	-1.129429	105.54
$tert$-C$_4$H$_9$NH$_2$	β-CH$_3$	-1.129094	-1.128887	105.37
$(CH_3)(C_2H_5)NH$	α-CH$_3$	-1.129923	-1.129834	105.67
	α-CH$_2$	-1.137179	-1.137090	107.95
	β-CH$_3$	-1.128947	-1.128858	105.36
$(C_2H_5)_2NH$	α-CH$_2$	-1.137195	-1.137124	107.96
	β-CH$_3$	-1.129042	-1.128972	105.40

[a]Weighted average for nonequivalent hydrogens attached to the same carbon.

Additional results are reported in Table 15.2 for selected NH, NN, CN, and CH bonds, and in Table 15.3 for CH bond energies in selected amines. These CH bond energies, and those of the CN and CC bonds given by (10.37), with charges deduced from NMR shifts [139], offer an estimate for $\Delta E_a^* \approx \sum_{k<l} \varepsilon_{kl}$. Comparison with similar calculations, but using the lump charge of all the hydrogens deduced from charge normalization [139], suggests that the CH bonds of Table 15.3 are (on the average) overestimated by ~ 0.4 kcal/mol for the primary amines, and seemingly correct for the secondary amines. The bottom line is that Eq. (10.15) and the results of Table 15.3 are reasonable estimates.

Table 15.1 offers pertinent information:

1. The result obtained for hydrazine indicates that $2 \times 100.08 = 200.16$ kcal/mol is the ΔE_a^* energy of each NH$_2$ group in the ground-state molecule. On the other hand, literature ΔH_f° data [285,286] and ZPE $+ H_T - H_0$ results [204] lead to $\Delta E_a^* = 181.06$ kcal/mol for the isolated NH$_2$ molecule (Table 12.3); hence RE $= 19.10$ kcal/mol, from Eq. (12.5).

2. Hence the CN dissociation energy deduced for methylamine from Eq. (12.1), namely, $D_{CN} = 92.00$ kcal/mol. Finally, Sanderson's approximation (12.14) leads to $\varepsilon_{CN} = 60.45$ kcal/mol for this molecule.

3. In a different approach, one obtains $\varepsilon_{CN} = 60.44$ kcal/mol for the CN bond found in methylamine, from its total atomization energy and the energies of its CH and NH bonds.

The CN Bond Energy Formula

The first thing to do is to learn how to write the bond energy formula [Eq. (10.37)] for carbon–nitrogen bonds

$$\varepsilon_{CN} = \varepsilon_{CN}^{\circ} + a_{CN}\Delta q_C + a_{NC}\Delta q_N \tag{15.1}$$

namely, to find the appropriate ε_{CN}° parameter. Methylamine, $CH_3 NH_2$, is selected as a convenient reference, with $\varepsilon_{CN} = 60.44\,kcal/mol$ as the sought-after reference intrinsic bond energy ε_{CN}°, of Eq. (15.1). Hence

$$\varepsilon_{CN} = 60.44 + a_{CN}\Delta q_C + a_{NC}\Delta q_N \ kcal/mol \tag{15.2}$$

where Δq_C and Δq_N are now expressed with respect to the carbon and nitrogen net charges, respectively, of methylamine.

Net Charges of the Nitrogen Atoms

Let us begin with trimethylamine. Its ^{13}C NMR shift, 47.56 ppm from TMS [165], gives $q_C = 28.92$ me. With $\varepsilon_{CH} = 104.96\,kcal/mol$ (Table 15.1), we deduce $q_H = -6.36_4$ me from the standard formula

$$\varepsilon_{CH} = 108.081 - 0.247q_C - 0.632q_H \ kcal/mol$$

and thus, from charge normalization, $q_N = -29.50$ me. (An uncertainty of ±1 me on q_C translates into an uncertainty of ±0.5 me on q_N.) The same calculation also indicates that $\varepsilon_{CN} = 71.34\,kcal/mol$ (Table 15.2). Next we compare the CN bond energies of the two molecules, methylamine and trimethylamine, with the help of Eq. (15.2), in kcal/mol:

$$60.44 = 71.34 - 0.603\left(q_C^{MeNH_2} - 28.92\right) - 0.448\left(q_N^{MeNH_2} + 29.50\right)$$

Using $\delta_C = 28.3$ ppm from TMS, we find the net charge of the methylamine carbon, $q_C^{MeNH_2} = 31.77$ me, and hence $q_N = -9.00$ me in the methylamine molecule. In similar fashion, with $q_C = 30.44$ me for the carbon of dimethylamine, we deduce from its CN bond energy (Table 15.2) that $q_N = -23.35$ me. This is as far as we can go with our present means. The internal consistency of these results is more important than their precise values, which could be refined, but energy calculations will dispense the verdict.

The solutions proposed for the nitrogen atoms are probably not as accurate as those given for the sp^3 carbon atoms. Still, they offer a valid basis for approximate but realistic energy calculations.

Selected Nitrogen–Nitrogen Bonds

A few typical nitrogen–nitrogen bonds are briefly examined. The key is in the result obtained from dimethylamine, leading to $RE(CH_3NH\bullet) = 9.19\,kcal/mol$. With ΔH_f° $(CH_3NH\bullet) = 43.6 \pm 3\,kcal/mol$ [246] and $ZPE + H_T - H_0 = 31.78\,kcal/mol$ [139], we get $\Delta E_a^* = 471.54\,kcal/mol$ for this radical; hence $D_{CN} = 89.27\,kcal/mol$ for dimethylamine, and thereby the reorganizational energy indicated above.

For methylhydrazine, one obtains $\Delta E_a^* = 725.50\,kcal/mol$ from $\Delta H_f^\circ = 22.6 \pm 0.1\,kcal/mol$ [246] and $ZPE + H_T - H_0 = 52.00\,kcal/mol$ [44]. Thus we deduce $D_{NN} = 72.90$ and $\varepsilon_{NN} = 44.61\,kcal/mol$.

For 1,2-dimethylhydrazine, it is now $\Delta E_a^* = 1014.94\,kcal/mol$ from $\Delta H_f^\circ = 22.0 \pm 1\,kcal/mol$ [246] and $ZPE + H_T - H_0 = 70.20\,kcal/mol$. Thus we deduce $D_{NN} = 71.86$ and $\varepsilon_{NN} = 53.48\,kcal/mol$.

Finally, for hydrazine itself the results are $D_{NN} = 74.70$ and $\varepsilon_{NN} = 36.30$ kcal/mol.

The methyl groups visibly inject electrons into the NN linkage, raising its intrinsic energy, but it is also clear that the relatively large reorganizational energies of the reaction products protect the nitrogen–nitrogen bonds in some way.

15.3 RESULTS

Straightforward applications of the theory are presented in the atom-by-atom approach, as exemplified in Table 15.4, using charges deduced from NMR shifts, Eq. (6.8) for the carbon atoms, and Eqs. (6.12)–(6.14) for the nitrogen atoms. (CN bond dissociation energies and comparisons with the corresponding intrinsic bond energies are described in Chapter 12 for both alkylamines and selected nitroalkanes.)

In the calculation of atomization energies, however, we meet with a difficulty. The hydrogen atoms, whose lump sums of atomic charges are deduced from charge normalization using the charges of the carbon and nitrogen atoms, are attached both to carbon and to nitrogen atoms, although the latter bonds concern only a small part of all the hydrogen atomic charges. Thus we can propose an approximate solution.

TABLE 15.4. Isopropylamine (me and kcal/mol)

Atom k	$\sum_l a_{kl}$	q_k (me)	$q_k \sum_l a_{kl}{}^a$
C_α	-1.826	29.60	-54.050
C_α	-1.229	32.04	-39.377
N	-0.842	-17.55	14.777
$H_{(N)}$	-0.794	1.30	-1.032
$H_{(C)}$	-0.632	-78.73	49.757

^{a}The sums $\sum_{k<l} \varepsilon_{kl}^\circ = 1242.63\,kcal/mol$ and $\sum_k \sum_l a_{kl} q_k$ give $\Delta E_a^* = 1172.30\,kcal/mol$, for $\Delta H_f^\circ = -19.87\,kcal/mol$.

TABLE 15.5. Standard Enthalpy of Formation of Amines (kcal/mol)

Molecule		ΔE_a^*	ZPE + ΔH	ΔH_f° Calculated[b]	ΔH_f° Experimental[a]
1	CH_3NH_2	581.37	41.48	−5.92	−5.50 ± 0.07
2	$C_2H_5NH_2$	876.22	59.41	−12.21	−11.35 ± 0.17
3	$n\text{-}C_3H_7NH_2$	1169.75	77.62	−16.90	−16.77 ± 0.13
4	$iso\text{-}C_3H_7NH_2$	1172.30	77.20	−19.87	−20.02 ± 0.17
5	$n\text{-}C_4H_9NH_2$	1463.29	95.83	−21.61	−22.0 ± 0.2
6	$iso\text{-}C_4H_9NH_2$	1464.64	95.49	−23.30	−23.57 ± 0.13
7	$sec\text{-}C_4H_9NH_2$	1465.60	95.03	−24.72	−25.4 ± 0.4
8	$tert\text{-}C_4H_9NH_2$	1469.32	95.21	−28.26	−28.90 ± 0.15
9	$cyclo\text{-}C_6H_{11}NH_2$	1928.18	118.89	−23.40	−25.07 ± 0.31
10	$NH_2(CH_2)_2NH_2$	1041.47	70.80	−3.94	−4.07 ± 0.14
11	$(CH_3)_2NH$	868.18	59.44	−4.14	−4.43 ± 0.12
12	$(CH_3)(C_2H_5)NH$	1163.56	77.36	−10.97	−11 ± 0.5
13	$(C_2H_5)_2NH$	1459.54	94.94	−18.75	−17.16 ± 0.31
14	$(n\text{-}C_3H_7)_2NH$	2046.11	131.72	−27.28	−27.84 ± 0.38
15	$(iso\text{-}C_3H_7)_2NH$	2051.57	130.94	−33.52	−34.4 ± 0.1
16	$(n\text{-}C_4H_9)_2NH$	2633.22	168.60	−36.26	−37.4 ± 0.3
17	$(iso\text{-}C_4H_9)_2NH$	2635.73	167.13	−40.07	−43.21 ± 0.65
18	$(CH_3)_3N$	1158.64	77.76	−5.66	−5.67 ± 0.18
19	$(CH_3)_2(C_2H_5)N$	1452.23	95.91	−10.47	−11
20	$(CH_3)(C_2H_5)_2N$	1747.26	113.00	−17.78	−17
21	$(CH_3)_2(tert\text{-}C_4H_9)N$	2042.06	131.78	−23.17	−21
22	$(C_2H_5)_3N$	2040.28	131.11	−22.06	−22.06 ± 0.19

[a]The results for **6, 9, 10, 14,** and **17** are reported in Ref. 247. Those of **1–4, 7, 11, 13, 18,** and **22** are from Ref. 248. For **5, 12, 15, 16,** and **19–21**, see Ref. 246. The enthalpies of **12** and **19–21** are estimated values. The ZPE + $H_T − H_0$ results are reported in Ref. 139. Original experimental data were used for **1** [249], **2–4** [287], **11** [288], **13** [289], and **18** [290], as cited in Ref. 44. The other results were obtained in the harmonic oscillator approximation using appropriately scaled fundamental frequencies deduced from B3LYP/6-311G** calculations [247], for $T = 298.15$ K. The scale factor of 0.96852 was determined by means of comparisons with the experimental frequencies of ammonia, methylamine, ethane, and propane. [b]The ^{13}C NMR shifts are from Ref. 165; those of ^{15}N are from Ref. 149. The rules of [165] were used for the ^{13}C shifts of $(CH_3)_2NH$ and $(CH_3)_2(C_2H_5)N$, together with the NMR results given in 166.

The charges of the amines considered here were calculated following a modified Del Re approach [34] that duplicates within ±0.35 me the carbon charges of alkanes (root-mean-square deviation) and within 0.75 and ∼1.38 me those of the carbon and nitrogen atoms of amines, which were deduced from NMR shifts. These results indicate that the charges of the hydrogens attached to nitrogen vary little, say, $q_H \simeq 1.3 ± 0.5$ me, and suggest a simple approximation; we shall carry out our calculations by letting $q_H = 1.30$ me, for an uncertainty of ∼0.12 kcal/mol.

The final results, presented in Table 15.5, are still reasonably accurate, as shown by the unsigned average error of 0.68 kcal/mol.

CHAPTER 16

OXYGEN–CONTAINING MOLECULES

16.1 ETHERS

A new reference bond energy must be introduced at this point: ε_{CO}° for the CO single bond of diethylether, selected as reference:

$$\varepsilon_{CO}^{\circ} = 79.78 \text{ kcal/mol for } q_{C\alpha}^{\circ} = 31.26 \text{ me and } q_O^{\circ} = 5.18 \text{ me}$$

The subscript α identifies the carbons adjacent to oxygen. These parameters yield the best fits with experimental energies [141]. Direct estimates using Eqs. (10.11) and (10.12) and the appropriate SCF potentials at the nuclei do indeed suggest $\varepsilon_{CO}^{\circ} \simeq 80$ kcal/mol [141]. The CC and CH bonds are treated in the usual manner, with reference to ε_{CC}° and ε_{CH}°, respectively. The corresponding $\varepsilon_{CO}^{o\prime}$ energy [Eq. (10.39)] is $\varepsilon_{CO}^{o\prime} = 104.635$ kcal/mol (Table 10.4).

The charges are deduced from Eqs. (6.8) and (6.16):

$$q_C = -0.148(\delta_C - \delta_C^{C_2H_6}) + 35.10 \text{ (me)} \quad C \neq \alpha$$

$$= -0.148(\delta_C - \delta_C^{Et_2O}) + 31.26 \text{ (me)} \quad C = C\alpha$$

$$q_O = -0.267(\delta_O - \delta_O^{Et_2O}) + 5.18 \text{ (me)}$$

The energy calculations are most conveniently made using the atom-by-atom approach, as illustrated by the examples given in Table 16.1. The ^{17}O and ^{13}C

Atomic Charges, Bond Properties, and Molecular Energies, by Sándor Fliszár
Copyright © 2009 John Wiley & Sons, Inc.

TABLE 16.1. Atom-by-Atom Calculation of Selected Ethers

Molecule	Atom k	δ_k	$\sum_l a_{kl}$	q_k (me)	$q_k \sum_l a_{kl}$
$CH_3OiC_3H_7$	C_{Me}	55.6	-1.453	32.78	-47.63
	C_α	72.9	-1.935	30.22	-58.48
	C_β	21.8	-1.229	32.73	-40.23
	O	-2.0	-1.002	7.45	-7.46
	H	—	-0.632	-135.91	85.90
				$\sum_{k<l} a_{kl}q_k$	-108.13
$iC_3H_7OtertC_4H_9$	C_α^{tb}	72.7	-2.176	30.25	-65.82
	C_β^{tb}	28.5	-1.229	31.74	-39.01
	C_α^{ip}	63.4	-1.935	31.63	-61.20
	C_β^{ip}	25.3	-1.229	32.21	-39.59
	O	62.5	-1.002	-9.77	9.79
	H	—	-0.632	-211.75	133.83
				$\sum_{k<l} a_{kl}q_k$	-179.61
$(sec\ C_4H_9)_2O$	C_α	74.3	-1.935	30.02	-58.09
	C_β	30.2	-1.470	31.49	-46.29
	C_γ	10.0	-1.229	34.48	-42.38
	$C_{\beta'}$	20.3	-1.229	32.95	-40.50
	O	41.5	-1.002	-4.17	4.17
	H	—	-0.632	-253.71	160.34
				$\sum_{k<l} a_{kl}q_k$	-210.01

NMR shift results of Ref. 140 were used, with $\delta_O^{Et_2O} = 6.5$ ppm from water and $\delta_C^{Et_2O} = 65.9$, and $\delta_C^{C_2H_6} = 5.8$ ppm from TMS.

Alternatively, one can use the general formula for dialkylethers [44]

$$\sum_k \sum_l a_{kl}\Delta q_k = A_1 \sum N_{CC}\delta_C + A_2 \sum \delta_C + A_3 \sum \delta_{C\alpha} + A_4 \delta_O$$

$$+ \Delta q_C^\circ \left(a_{CC} \sum N_{C\alpha C} + a_{CH} \sum N_{C\alpha H} - 2a_{HC} \right)$$

$$+ (n_C - 2)a_{HC}q_H^\circ - a_{HC}q_O^\circ$$

which incorporates the appropriate charge–NMR shift relationships. N_{CC} is the number of CC bonds formed by the carbon whose shift is δ_C. The $N_{CC}\delta_C$ term includes both $\delta_{C\alpha}$ (relative to the diethylether C_α atom) and δ_C (from ethane). $N_{C\alpha C}$ and $N_{C\alpha H}$ are, respectively, the number of CC and CH bonds formed by the C_α carbons. Moreover, $A_1 = a_{CC} - a_{CH}$; $A_2 = 4a_{CH} - a_{HC}$; $A_3 = 3a_{CH} + a_{CO} - a_{HC}$ (see Table 10.4); $\Delta q_C^\circ = q_{C\alpha}^\circ - q_C^{C_2H_6} = -3.84$ me and $q_O^\circ = 5.18$ me. Because of the large variations of q_O, one may use $A_4 = 0.1007$ kcal mol^{-1} ppm^{-1} for $a_{OC} = -0.804$ au ($\delta_O < 0$) or $A_4 = 0.0994$ for $a_{OC} = -0.800$ au ($\delta_O > 0$).

The rms deviation between calculated and experimental enthalpies of formation (see Table 16.2), 0.42 kcal/mol, illustrates the overall consistency of our charge

TABLE 16.2. Standard Enthalpy of Formation of Ethers[a] (kcal/mol)

Molecule		$\sum \varepsilon_{kl}$	E_{nb}	ZPE + ΔH	ΔH_f° Calculated	ΔH_f° Experimental[b]
1	$(CH_3)_2O$	796.09	−0.05	52.55	−44.06	−43.99 ± 0.12
2	$CH_3OC_2H_5$	1093.23	−0.09	70.19	−51.97	−51.72 ± 0.16
3	$CH_3O\text{-}n\text{-}C_3H_7$	1386.27	−0.25	87.83	−56.91	−56.82 ± 0.26
4	$CH_3O\text{-}iso\text{-}C_3H_7$	1389.74	−0.12	87.83	−60.25	−60.24 ± 0.23
5	$(C_2H_5)_2O$	1389.37	−0.17	87.83	−59.93	−60.26 ± 0.19
6	$CH_3O\text{-}n\text{-}C_4H_9$	1679.72	−0.37	105.47	−62.21	−61.68 ± 0.27
7	$CH_3O\text{-}tert\text{-}C_4H_9$	1684.26	−0.17	105.15	−66.87	−67.68 ± 0.31
8	$C_2H_5O\text{-}n\text{-}C_3H_7$	1682.34	−0.32	105.47	−64.78	−65.05 ± 0.30
9	$C_2H_5O\text{-}t\text{-}C_4H_9$	1980.86	−0.30	123.11	−75.01	−75.0 ± 0.5
10	$(n\text{-}C_3H_7)_2O$	1975.34	−0.47	123.11	−69.66	−69.85 ± 0.40
11	$(iso\text{-}C_3H_7)_2O$	1982.39	−0.29	123.11	−76.53	−76.20 ± 0.54
12	$i\text{-}C_3H_7Ot\text{-}C_4H_9$	2278.41	−0.70	140.75	−84.70	−85.5 ± 1.2
13	$(n\text{-}C_4H_9)_2O$	2561.87	−0.73	158.39	−78.92	−79.82 ± 0.27
14	$(sec\text{-}C_4H_9)_2O$	2568.14	−0.58	158.39	−86.04	−86.26 ± 0.41
15	Tetrahydropyran	1557.13	−0.49	92.87	−53.22	−53.39 ± 0.24
16	1,4-Dioxane	1352.45	−0.29	78.48	−75.63	−75.65 ± 0.22

[a]The approximation ZPE + ΔH (=ZPE +$H_T − H_0$) = 52.55 + 17.64(n_C − 2) kcal/mol [44], based on observed frequencies [291,292], was used for **1–14**, where n_C is the number of carbon atoms. For **15** and **16**, see Ref. 27. The nonbonded parts, E_{nb} [Eq. (10.3)], are reported in Ref. 35.
[b]The experimental values are reported in Ref. 248, except those of **6–8** described in Ref. 293 and that of **9** given in Ref. 294. For **7**, the result $\Delta H_f^\circ = −67.45 ± 0.45$ kcal/mol is reported in Ref. 295. That of **16** is from Ref. 296. The ^{13}C and ^{17}O NMR shifts are from Ref. 140.

and energy calculations. Di-*tert*-butylether, however, fails in that respect. Using the ^{13}C and ^{17}O shifts measured by Delseth et al. [140], one obtains $\Delta H_f^\circ = −95.6$ kcal/mol for this molecule, which is in error by 8.5 kcal/mol with respect to the experimental result, $\Delta H_f^\circ = −87.1 ± 0.4$ kcal/mol. But an anomaly is detected in the ^{17}O charge–shift correlation (Fig. 6.9), suggesting that Eq. (6.16) would predict too negative an oxygen atom. Thus we suspect that something fails with the use of the Delseth results for this molecule in Eqs. (6.8) and (6.16) to get charge results. Indeed, the use of calculated charges leads to the correct result, $\Delta H_f^\circ = −87.45$ kcal/mol [35].

16.2 ALCOHOLS

Unfortunately, we cannot benefit from an a priori knowledge of the oxygen net charges to assist our search. But a posteriori, with the ^{17}O NMR results given in Ref. 297, it appears appropriate to write

$$q_O \approx −0.165\delta_O^{H_2O} + 4.85 \text{ me} \qquad (16.1)$$

TABLE 16.3. 2-Butanol (kcal/mol)

Atom k	δ_k^a	$\sum_l a_{kl}$	q_k (me)	$q_k \sum_l a_{kl}$
C-1	22.9	−1.229	32.540	−39.992
C-2	69.0	−1.935	30.801	−59.600
C-3	32.3	−1.470	31.148	−45.788
C-4	10.2	−1.229	34.419	−42.301
O	34.0	−1.049	−0.760	0.797
H	—	−0.632	−128.148	80.990

[a]From Refs. 138 (^{13}C) and 297 for the ^{17}O shifts. The ^{13}C shifts are from TMS; that of ^{17}O is from water.

where $\delta_O^{H_2O}$ is the ^{17}O NMR shift relative to external water. This formula is advocated by applications to energy calculations and comparisons with experimental results. The reference charge, 4.85 me, is roughly estimated within ± 0.5 me, along with its corresponding OH reference bond energy, $\varepsilon_{OH}^{o\prime} = 115.678$ kcal/mol. It is understood that this parameterization is tentative and amenable to future improvements; the charges of the hydroxyl groups are not as firmly established as are those of the carbon atoms. Energy calculations prove nonetheless satisfactory. They reveal

TABLE 16.4. Standard Enthalpy of Formation of Alcohols^{a,b} (kcal/mol)

	Molecule	$\sum \varepsilon_{kl}$	E_{nb}	ZPE + ΔH	ΔH_f° Calculated	ΔH_f° Experimental^c
1	H_2O	232.33		15.25	−57.80	−57.80 ± 0.04
2	CH_3OH	512.31	−0.02	33.69	−48.74	−48.07 ± 0.05
3	C_2H_5OH	808.57	−0.13	51.69	−56.48	−56.24 ± 0.07
4	n-C_3H_7OH	1101.68	−0.28	69.76	−61.04	−61.17 ± 0.30
5	iso-C_3H_7OH	1105.81	−0.23	69.39	−65.49	−65.12 ± 0.13
6	n-C_4H_9OH	1395.17	−0.41	87.87	−65.92	−65.79 ± 0.14
7	iso-C_4H_9OH	1396.57	−0.34	87.55	−67.58	−67.84 ± 0.21
8	sec-C_4H_9OH	1398.82	−0.30	87.46	−69.88	−69.98 ± 0.23
9	$tert$-C_4H_9OH	1403.38	−0.29	86.89	−74.99	−74.72 ± 0.21
10	1-$C_5H_{11}OH$	1688.81	−0.54	105.91	−71.03	−70.66 ± 0.18
11	2-$C_5H_{11}OH$	1692.55	−0.46	105.61	−74.99	−75.18 ± 0.36
12	3-$C_5H_{11}OH$	1691.28	−0.46	(105.61)	−73.72	−75.21 ± 0.27
13	2-Me-1-Butanol	1689.21	−0.50	105.69	−71.61	−72.19 ± 0.35
14	3-Me-1-Butanol	1690.41	−0.39	105.73	−72.66	−72.02 ± 0.35
15	2-Me-2-Butanol	1695.44	−0.36	(105.0)	−78.39	−79.07 ± 0.35
16	$HOCH_2CH_2OH$	904.89	−0.20	55.43	−91.07	−92.7 ± 0.4
17	$cyclo$-$C_6H_{11}OH$	1861.01	−0.75	(111.5)	−68.44	−68.38 ± 0.42

[a]The nonbonded contributions E_{nb} [Eq. (10.3)] are from Ref. 35.
[b]The results for ZPE + ΔH (=ZPE + $H_T - H_0$) were obtained in the harmonic oscillator approximation from appropriately scaled fundamental frequencies deduced from B3LYP/6-31G** calculations [247], for $T = 298.15$ K. The result for **12** is based on that of **11**; that of **15** is inferred from 3-Me-2-butanol considering that in the presence of $tert$-butyl-like structures the ZPE + $H_T - H_0$ energy is lowered by $\sim$0.3–0.4 kcal/mol [44].
[c]The experimental values are reported in Ref. 248.

that $(\partial E/\partial N)_O^\circ$ should be taken at ~ -0.8 for hydroxyl oxygen atoms in the evaluation of a_{OH} and a_{OC} [Eq. (10.41)]. The energies were calculated in the atom-by-atom approach (as in the example given in Table 16.3) and are reported in Table 16.4. The carbons are calculated as shown for the ethers, using, namely, $q_C = -0.148(\delta_C - \delta_C^{Et_2O}) +$ 31.26 me for the α-carbons and $q_C = -0.148(\delta_C - \delta_C^{C_2H_6}) + 35.1$ me otherwise.

A drastic approximation had to be made for the hydrogens; their total charge was deduced from charge normalization and was entirely treated as atomic charges of hydrogens attached to carbon, i.e. as if q_H was always null in the OH part, which is wrong, of course.

This shortcut does not help improve the quality of our presentation. Indeed, one single millielectron erroneously attributed to a hydrogen atom attached to carbon (with $a_{HC} = -0.632$), rather than to a hydroxyl H atom (with $a_{HO} = -1.000$ kcal mol^{-1} me^{-1}), renders $\sum_k \sum_l a_{kl} q_k$ too negative by 0.368 kcal/mol and lowers ΔE_a^* by that amount, which renders ΔH_f° less negative than what it really should be. The results of Table 16.4 suggest that the hydroxyl hydrogen of methanol should have less, and that of 3-$C_5H_{11}OH$ should have more, electronic charge than what is produced by our brute-force approximation. These circumstances, and the approximate nature of Eq. (16.1), are in part responsible for the unusually high errors made for some alcohols, averaging ~ 0.51 kcal/mol. Intramolecular hydrogen bonding should probably be considered in **16**.

The case of water is treated differently. It is correctly calculated as

$$\varepsilon_{OH} = \varepsilon_{OH}^{o\prime} + a_{OH} q_O + a_{HO} q_H$$
$$= \varepsilon_{OH}^{o\prime} - 0.400 q_O - 1.000\, q_H$$
$$= \varepsilon_{OH}^{o\prime} + 0.100\, q_O$$

so that Eq. (16.1) gives $\varepsilon_{OH} = 116.163$ kcal/mol and thus the results quoted in Table 16.4.

16.3 CARBONYL COMPOUNDS

The CO double bond of acetone introduces a new reference energy:

$$\varepsilon_{C=O}^\circ = 179.40 \text{ kcal/mol} \quad \text{for} \quad q_{C\alpha}^\circ = 14.0 \text{ me} \quad \text{and} \quad q_O^\circ = -21.2 \text{ me}$$

Direct estimates using the appropriate SCF potentials at the nuclei suggest $\varepsilon_{C=O}^\circ \simeq 181$ kcal/mol [141]. The CC and CH parameters are treated the usual way.

A general formula is developed as follows. The hydrogen charge variations are expressed relative to $q_H^\circ = -11.7$ me. Charge normalization, $\sum q_H = -(\sum q_C + \sum q_{C\alpha} + \sum q_O)$, gives

$$\Delta q_H = -\left(\sum \Delta q_C + \sum \Delta q_{C\alpha} + \sum \Delta q_O \right)$$
$$- \left(n_C q_C^{\circ C_2H_6} + n_{C\alpha} q_{C\alpha}^\circ + n_O q_O^\circ + n_H q_H^\circ \right) \tag{16.2}$$

where $n_{C\alpha}$, n_O, n_C, and n_H are the numbers of C atoms adjacent to O, of the O atoms, the C atoms not bonded to O, and of the H atoms, respectively. (This formula is also valid for dialkylethers.)

Summation of the $a_{kl}\Delta q_k$ terms and Eq. (16.2) gives the following general formula describing carbonyl compounds RR'CO with R = alkyl and R' = alkyl or H

$$\sum_k \sum_l a_{kl}\Delta q_k = A_1 \sum N_{CC}\delta_C + A_2 \sum \delta_C + A_3'\delta_{C\alpha} + A_4'\delta_O$$

$$+ \Delta q_{C\alpha}^{\circ}(a_{C\alpha C}N_{C\alpha C} + a_{C\alpha H}N_{C\alpha H} - a_{HC})$$
$$+ n_C a_{HC}q_H^{\circ} - a_{HC}q_O^{\circ} \qquad (16.3)$$

where

$$A_3' = 2a_{C\alpha H} + a_{CO} - a_{HC}$$
$$A_4' = a_{OC} - a_{HC}$$

The other parameters of (16.3) are those explained for the dialkylethers. $N_{CC}\delta_C$ includes both $\delta_{C\alpha}$ (from the acetone carbonyl-C atom) and δ_C (from ethane) for the atoms not bonded to O. Selected a_{kl} values are $a_{C\alpha H}(1.08) = -0.276$ and $a_{CO}(1.22) = -1.182$ au. Hence $A_3' = -0.727$ au. Assuming $\Delta q_C = -0.148\delta_C$ me, we get $A_3' = 0.0675$ kcal mol^{-1} ppm^{-1}.

Noting that $N_{C\alpha C} + N_{C\alpha H} = 2$, we rewrite Eq. (16.3) as follows

$$\sum_k \sum_l a_{kl}\Delta q_k = 0.0356 \sum N_{CC}\delta_C + 0.0529 \sum \delta_C + 0.0675\delta_{C\alpha} + A_4'\delta_O$$

$$+ 7.393 n_C + 5.07 N_{C\alpha C} - 19.42 \text{ kcal/mol} \qquad (16.4)$$

where n_C is the total number of C atoms. Here we cannot take a_{OC} as a constant because of the large variations of Δq_O. Using the γ data of Table 10.2 and the appropriate first and second derivatives, $(\partial E_O/\partial N_O)^{\circ} = -0.316$ and $(\partial^2 E_O/\partial N_O^2)^{\circ} = 0.50$ au, respectively, one obtains from Eq. (10.32) that $a_{OC}(1.22) = -1.065 - 0.254 \times \Delta q_O$ and $A_4' = -(0.058 + 0.254\Delta q_O)$. Using now $\Delta q_O(\text{carbonyl}) \simeq 2.7\delta_O$ me [Eq. (6.17)], we get

$$A_4' \simeq -(0.098 + 1.16 \times 10^{-3} \times \delta_O) \text{ kcal mol}^{-1} \text{ ppm}^{-1}$$

The validity of this approximation is best illustrated by the results offered in Table 16.5. Although Eq. (16.4) permits accurate calculations of atomization energies and represents a simple and valuable tool, it must be made clear that the charge-NMR shift correlations used for the atoms of the carbonyl group (i.e., those involved in the $A_3'\delta_{C\alpha}$ and $A_4'\delta_O$ terms) are empirical. On the other hand, it turns out that for the ketones $A_3'\delta_{C\alpha} + A_4'\delta_O$ amounts to less than 5% of $\sum_k \sum_l a_{kl}\Delta q_k$. Hence, with some reservations in mind, it seems reasonable to claim that at least the main features of the theory underlying these calculations withstand challenging tests like those presented in Table 16.5 [44,141].

TABLE 16.5. Comparison between Calculated and Experimental Enthalpies of Formation of Selected Carbonyl Compounds (kcal/mol)

	Molecule	$\sum \varepsilon_{kl}$	E_{nb}	ZPE + ΔH	Calculated[a]	Experimental[b]
					\multicolumn{2}{c}{ΔH_f°}
1	CH_3CHO	675.48	-0.30	36.65	-39.81	-39.73 ± 0.12
2	C_2H_5CHO	969.67	-0.36	54.95	-45.14	-45.45 ± 0.21
3	n-C_3H_7CHO	1262.01	-0.75	73.25	-48.94	-48.98 ± 0.34
4	$(CH_3)_2CO$	976.35	-0.07	54.59	-51.89	-51.90 ± 0.12
5	$CH_3COC_2H_5$	1269.94	-0.45	72.89	-56.93	-57.02 ± 0.20
6	CH_3COn-C_3H_7	1563.52	-0.42	91.19	-61.55	-61.92 ± 0.26
7	CH_3COi-C_3H_7	1564.75	-0.55	91.19	-62.92	-62.76 ± 0.21
8	$(C_2H_5)_2CO$	1563.42	-0.92	91.52	-61.63	-61.65 ± 0.21
9	CH_3COn-C_4H_9	1858.07	-0.55	109.49	-67.31	-66.96 ± 0.21
10	CH_3COt-C_4H_9	1860.69	-0.6	109.5	-70.0	-69.28 ± 0.25
11	C_2H_5COn-C_3H_7	1857.36	-1.05	109.49	-67.10	-66.51 ± 0.22
12	C_2H_5COi-C_3H_7	1858.65	-1.05	109.49	-68.39	-68.38 ± 0.27
13	C_2H_5COt-C_4H_9	2153.99	-1.2	128.12	-74.62	-74.99 ± 0.33
14	$(i$-$C_3H_7)_2CO$	2153.10	-1.85	128.12	-74.38	-74.40 ± 0.28

[a]The ^{13}C NMR spectra are from Refs. 138 and 142; those of the ^{17}O atoms are given in Ref. 142.
[b]The experimental values are reported in Ref. 248.

With the $\sum_k \sum_l a_{kl}\Delta q_k$ expressed as indicated in Eq. (16.3), one must use the ε_{kl}° bond energy references of Eq. (10.37), namely, in kcal/mol, $\varepsilon_{CC}^\circ = 69.633$, $\varepsilon_{CH}^\circ = 106.806$ (as in ethane), and $\varepsilon_{C=O}^\circ = 179.40$ (like the CO bond in acetone).

The nonbonded contributions are taken from [44,141]; for carbonyl compounds, they should be considered with some reservations and are used here on a tentative basis. As concerns ZPE + $H_T - H_0$, an increment of 18.3 kcal/mol was assumed for each added CH_2 group with respect to the closest "parent" compound [27,44,141,180].

It seems fair to conclude that the charges used in these calculations, based on ^{13}C and ^{17}O NMR shifts, are adequate; the unsigned average deviation between calculated and experimental enthalpies of formation, 0.22 kcal/mol, certainly supports this view.

CHAPTER 17

PERSPECTIVES

This brings us to the question: and now?

The physics underlying this bond energy theory is attractive for the concepts it carries into effect: the role of the electronic charge at the bond-forming atoms. Abundant proof has taught us that given a good knowledge of the charge distribution in a molecule (e.g., by means of ^{13}C, ^{15}N, ^{17}O NMR chemical shift results), calculations are not only extremely simple and quick but also very accurate for a variety of both saturated and unsaturated molecules. So much for the past.

But difficulties exist. Although—so they say—everything started with a big bang, it better be clear in our minds that Nature operates in delicate manner, with little effort; small, sometimes very small causes (in our perception of life) have important consequences. We should remember this rule of thumb: a charge variation of one sole millielectron at the ends of a sigma bond translates into a change in the order of one kcal/mol in bond energy, which is a lot. Indeed, we now know that the charge distributions of alkanes that correlate with carbon NMR shifts, adiabatic ionization potentials, and molecular atomization energies are precisely those that satisfy the so-called inductive effects by allowing the smallest possible variations. In other words, maximum precision in the evaluation of atomic charges is required for our intended calculations of bond energies. Simply said, most accurate tools are a must.

Does this imply that it commands the use of complicated means? No! Not necessarily. And this work should not be taken as the end of the affair. Quite on the contrary, it is just the beginning—improvements are certainly upon us.

Atomic Charges, Bond Properties, and Molecular Energies, by Sándor Fliszár
Copyright © 2009 John Wiley & Sons, Inc.

One could perhaps think of an approach consisting of straightforward calculations of the NMR shifts of the heavy atoms, a task that can be performed with accuracy thanks to methods developed by a number of authors [165,169,250,298]. Evidently, this would still leave us with the problem of assigning the charges of the hydrogen atoms attached to the various atoms, paying particular attention to those attached to atoms other than carbon; but that problem could surely be overcome. Alternative examples exist for sigma systems [34–37]; they could perhaps offer useful hints, but we prefer to keep an open mind on that matter and try harder for simpler solutions.

Still, we strongly feel that a patient processing of general theory, rather than an accumulation of empirically generated data, should offer valid solutions permitting to extend the present methods to cover most, if not all, of organic chemistry.

This work has also taught us—and this is a promising new field of applications—how to get a fresh insight into individual bond dissociation energies and their dependence on whatever events modify the electronic structures of the concerned reaction sites. This area only awaits the inventiveness of our community, a community that never lacks questions and ideas to get the answers. So much for the future.

Sure, in most instances simplifying assumptions cannot be proven a priori to be of minor import, but this work is the story of a rough simple approximation in the translation of theory into models, rules, and properties that keep chemistry in the hands of chemists. The concept of bond energies that depend on the charges of the bond-forming atoms is not new, but the simple calculation of the required charges surely calls for sustained inventive attention in the future.

All is quantum chemistry, but quantum chemistry is not all.

Goodbye, Dear Reader!

APPENDIX: WORKING FORMULAS

A compendium of final working formulas is offered here with the intent to facilitate numerical applications. All these formulas and the appropriate numerical parameters are described in the text; this presentation simply inserts the parameters as required, as they were used in our calculations.

A.1 CHARGE–NMR SHIFT CORRELATIONS

For any sp^3 *carbon bonded to carbon and hydrogen*, use

$$q_C = -0.148(\delta_C - \delta_C^{C_2H_6}) + 35.1 \quad \text{me} \tag{A.1}$$

where $\delta_C - \delta_C^{C_2H_6}$ is the NMR shift relative to ethane. This formula also applies to the carbon–nitrogen bonds of amines, but not to sp^3 carbons attached to oxygen, as oxygen introduces an "extra" downfield shift at its bonded α-carbon. For the sp^3 carbons of *ethers and alcohols* attached to oxygen, use

$$q_C = -0.148(\delta_C - 65.9) + 31.26 \quad \text{me} \tag{A.2}$$

where δ_C is the carbon shift from TMS and 65.9 ppm is the carbon NMR shift of diethylether, relative to TMS; the carbon net charge of that reference is 31.26 me. For *aldehydes and ketones*, use

$$q_C = -0.148(\delta_C - 204.9) + 14.0 \quad \text{me} \tag{A.3}$$

Atomic Charges, Bond Properties, and Molecular Energies, by Sándor Fliszár
Copyright © 2009 John Wiley & Sons, Inc.

for the carbon attached to oxygen; $\delta_C = 204.9$ ppm is for acetone, from TMS, with a carbonyl–carbon net charge of 14.0 me. For *olefinic sp^2 carbons*, use

$$q_C = 0.15(\delta_C - 122.8) + 7.7 \quad \text{me} \tag{A.4}$$

where δ_C is expressed relative to tetramethylsilane (TMS), and $\delta_C^{C_2 H_4} = 122.8$ ppm.

For the *benzenoid sp^2 carbons*, one must consider the variations of both the σ and π electronic charges, $\Delta q_\sigma = m\Delta q_\pi$, so that $\Delta q_C = \Delta q_\sigma + \Delta q_\pi = (m + 1)\Delta q_\pi$, with $\Delta\delta_C = 157\Delta q_\pi$; hence

$$\Delta\delta_C = \frac{157}{m + 1}\Delta q_C \tag{A.5}$$

The slope (157) is expressed here in ppm/electron, with Δq_π and Δq_C in electron units. Reverting now to millielectron units, we get

$$q_C = 1.2(\delta_C - \delta_C^{C_6 H_6}) + 13.2 \quad \text{me} \tag{A.6}$$

for any benzenoid carbon bonded *only* to other benzenoid carbons, with $m = -0.814$, whereas

$$q_C = 0.637(\delta_C - \delta_C^{C_6 H_6}) + 13.2 \quad \text{me} \tag{A.7}$$

applies to substituted aromatic carbon atoms, with $m = -0.90$, such as those found in toluene, styrene, or biphenyl, for example.

The *N-15 resonance shifts of alkylamines and nitroalkanes* are expressed in ppm from HNO_3, in methanol. Separate correlation lines are observed for mono-, di- and trialkylamines, namely

$$q_N = 0.218(\delta_N - 371.1) - 9.00 \quad \text{me (primary amines)} \tag{A.8}$$
$$q_N = 0.247(\delta_N - 363.3) - 23.35 \quad \text{me (secondary amines)} \tag{A.9}$$
$$q_N = 0.168(\delta_N - 356.9) - 29.50 \quad \text{me (tertiary amines)} \tag{A.10}$$

where 371.1 ppm is the shift of the methylamine nitrogen, and -9.00 me its net charge; 363.3 ppm is the shift observed for dimethylamine, with $q_N = -23.35$ me, and 356.9 ppm is that of trimethylamine, with $q_N = -29.50$ me. For *nitroalkanes*, the formula is

$$q_N = 0.253(\delta_N - \delta_N^{CH_3 NO_2}) + q_N^{CH_3 NO_2} \tag{A.11}$$

where the nitrogen net charge in nitromethane ($q_N^{CH_3 NO_2}$) has yet to be determined.

The correlations involving O-17 atoms concern dialkylethers, alcohols, aldehydes, and ketones. For the *dialkylether oxygen atoms*, the appropriate formula is

$$q_O = -0.267(\delta_O - 6.5) + 5.18 \quad \text{me} \tag{A.12}$$

where 5.18 me is the assumed oxygen net charge in diethylether and 6.5 ppm is its ^{17}O shift from water. For *alcohols*, we use

$$q_O \approx -0.165(\delta_O - \delta_O^{H_2O}) + 4.85 \quad \text{me} \tag{A.13}$$

where $\delta_O - \delta_O^{H_2O}$ is the ^{17}O NMR shift relative to external water. For the oxygens occurring in *carbonyl groups*, our tentative formula is

$$q_O \approx 2.7(\delta_O - \delta_O^{acetone}) - 21.2 \quad \text{me} \tag{A.14}$$

A.2 GENERAL ENERGY FORMULAS

For the *normal and branched alkanes* as well as for *six-membered chair, boat, and twist–boat cycloalkanes*, the following formula for the energy of atomization, ΔE_a^* (in kcal/mol), includes the nonbonded interactions in an approximate, but very accurate manner:

$$\Delta E_a^* \simeq 710.54(1 - m) + 290.812(n - 2 + 2m)$$
$$+ 0.03244 \sum_k N_{C_kC} \delta_{C_k} + 0.05728 \sum_k \delta_{C_k} + 0.822 n_g \tag{A.15}$$

where n = number of carbon atoms, m = number of cycles, δ_{C_k} = carbon NMR shift *relative to ethane*, N_{C_kC} = number of carbon–carbon bonds formed by atom C_k, and n_g = number of *gauche* interactions (such as exemplified by one of the methyl protons of the axial methylcyclohexane, for instance, interacting with the axial protons at C-3 and C-5 of the cyclohexane ring).

Standard enthalpies of formation, ΔH_f° (gas, 298.15K), calculated along these lines, using charges inferred from NMR chemical shifts and the standard formula

$$\Delta E_a^* = \sum_k n_k \left[\Delta H_f^\circ(A_k) - \frac{5}{2}RT \right] + ZPE + (H_T - H_0) - \Delta H_f^\circ \tag{A.16}$$

agree within 0.24 kcal/mol (root-mean-square deviation) with their experimental counterparts. The enthalphy of formation ΔH_f° is thus[1]

$$\Delta H_f^\circ = ZPE + H_T - H_0 - 20.185n - 27.698(1 - m)$$
$$- 0.03244 \sum_k N_{C_kC} \delta_{C_k} - 0.05728 \sum_k \delta_{C_k} - 0.822 n_g \tag{A.17}$$

Taking advantage of the remarkable additive properties of the $ZPE + H_T - H_0$ energies, it also becomes possible to integrate them in simple formulas for the calculation

---

[1]Smaller or larger cycles are not described by this formula; their appropriate charge–NMR shift correlations are yet to be investigated in a systematic fashion, as well as other possible effects that may be due to changes in hybridization.

of $\Delta H_\mathrm{f}^\circ$, namely, for *linear and branched paraffins*, $C_n H_{2n+2}$, we obtain

$$\Delta H_\mathrm{f}^\circ = -\Bigg(16.219 + 1.972n + 0.343 n_\mathrm{br} + 0.03244 \sum_k N_{C_kC} \delta_{C_k}$$

$$+ 0.05728 \sum_k \delta_{C_k} \Bigg) \qquad (A.18)$$

where n_br is the number of branchings. Situations involving an extreme steric crowding (as with two *tert*-butyl groups attached to a same carbon) are not well represented by this formula. (The suspected reasons are a possible breakdown of the simple scheme for evaluating $ZPE + H_T - H_0$ energies and/or a failure of the charge–NMR shift relationship due to slight changes in hybridization.)

The undisputed success of these very simple and accurate formulas, in which the $0.03244 \sum_k N_{C_kC} \delta_{C_k} + 0.05728 \sum_k \delta_{C_k}$ part most adequately takes care of all deviations with respect to what would result from a simple bond additivity scheme involving assumed constant bond energy terms, surely illustrates the general validity of the approach rooted in charge-dependent bond energy terms in the first place, and of the concurrent theoretical parameterization of the method on top of it.

The same holds true for *olefins*, $C_n H_{2n}$, whose atomization energies

$$\Delta E_\mathrm{a}^* = \sum_{k<l} \varepsilon_{kl}^\circ + \sum_k \sum_l a_{kl} \Delta q_k + F \qquad (A.19)$$

are calculated using $\varepsilon_{CC}^\circ = 69.633$, $\varepsilon_{CH}^\circ = 106.806$ and $\varepsilon_{C=C}^\circ = 139.37\,\mathrm{kcal/mol}$, which are the references for ethane and ethylene, respectively. The point is that all the effects due to the shifts of the π-orbital centroids along the ethylenic double bond are duly accounted for by means of the appropriate transferable theoretical corrections, in the following general formulas, also using the theoretical a_{kl} parameters, namely, in kcal/mol units

$$\sum_k \sum_l a_{kl} \Delta q_k \simeq 0.0356 \sum N_{CC} \delta_C + 0.0529 \sum \delta_C + 0.18 \sum \delta_{C'}$$

$$+ 7.393n + 4.19 \sum N_{C'C} - 19.16 \qquad (A.20)$$

for *ethylene, 1-alkenes, trans-alkenes*, and *tetramethylethylene*, and

$$\sum_k \sum_l a_{kl} \Delta q_k \simeq 0.0356 \sum N_{CC} \delta_C + 0.0529 \sum \delta_C + 0.18 \sum \delta_{C'}$$

$$+ 7.393n + 4.00 \sum N_{C'C} - 18.22 \qquad (A.21)$$

for *cis-olefins*. Note that the ^{13}C NMR shifts of the sp^3 carbons δ_C are relative to ethane, while those of the sp^2 carbons $\delta_{C'}$ are from ethylene. N_{CC} is the number of carbon–carbon bonds formed by an sp^3 carbon with any other carbon atom, while $N_{C'C}$ is the number of bonds formed by an sp^2 carbon (marked C') with sp^3 carbon atoms.

The term in $\sum N_{C'C}$ indicates an a priori gain in stability of $\sim 4\,\mathrm{kcal/mol}$ for each CC bond formed by an sp^2 carbon, although this gain is somewhat counteracted by electronic charge redistributions. For example, in going from 2-methyl-1-pentene ($\sum N_{C'C} = 2$) to 3-methyl-1-pentene ($\sum N_{C'C} = 1$), ΔE_a^* does not decrease by ~ 4, but by only $2.6\,\mathrm{kcal/mol}$, but the leading term is clearly determined by $\sim 4\sum N_{C'C}$. On these grounds it is easy to predict that methylenecyclohexane (with an exocyclic double bond and $\sum N_{C'C} = 2$ and $\Delta H_f^\circ = -7.2\,\mathrm{kcal/mol}$) should be less stable than 1-methylcyclohexene ($\sum N_{C'C} = 3$, $\Delta H_f^\circ = -10\,\mathrm{kcal/mol}$). This rationale is relevant in the discussion of the relative thermochemical stabilities of exo- vs. endocyclic double bonds. Of course, the fact alone that a double bond is exocyclic does not necessarily imply a loss in stability. For example, ethylidenecyclohexane [ΔH_f° (gas) $= -14.7\,\mathrm{kcal/mol}$] and 1-ethylcyclohexene ($\Delta H_f^\circ = -15.0\,\mathrm{kcal/mol}$), for which $\sum N_{C'C} = 3$ in both cases, are visibly similar in terms of their thermochemical stabilities. These theoretical expectations, based on the $\sim 4\sum N_{C'C}$ term, fully confirm similar views expressed by Fuchs and Peacock [299] suggesting that the cyclic compounds differ in double-bond substitution rather than conformational stability—an area certainly worth additional investigations. The same trends are observed for the five-membered ring analogs, as revealed by their heats of formation (gas, 298.15K, indicated in kcal/mol). Indeed, 1-methylcyclopentene (-0.86), with $\sum N_{C'C} = 3$, is more stable than methylenecyclopentane (2.4), with $\sum N_{C'C} = 2$. In turn, 1-ethylcyclopentene (-4.72) and ethylidenecyclopentane (-4.33), both with $\sum N_{C'C} = 3$, are quite similar in stability. This similarity between five- and six-membered cyclic hydrocarbons is not surprising, but it is not any longer observed for the three-membered ring analogs, probably because a considerable ring strain introduced by short endocyclic double bonds.

Dienes, Allenes, Alkynes and Benzene

So far, no general formulas have been worked out for these hydrocarbons, thus precluding the evaluation of the $\sum_k \sum_l a_{kl}\Delta q_k$ terms for the allenes and alkynes, but the results obtained for the alkanes and the alkenes offer interesting clues that permit rough but (under the circumstances) acceptable back-of-the-envelope estimates: we gain in "chemical understanding" what is lost in precision. The argument rests on two circumstances:

- The selection of $\varepsilon_{CH}^\circ = 106.806$ and $\varepsilon_{CC}^\circ = 69.633\,\mathrm{kcal/mol}$ as common bond energy references and of $\varepsilon_{C=C}^\circ = 139.37\,\mathrm{kcal/mol}$ for the double bonds
- The fact that the $\sum_k \sum_l a_{kl}\Delta q_k$ terms are quite similar for the alkanes and the alkenes having the same number of carbon atoms, except for ethane versus ethylene

If we decide not to be too critical about discrepencies of a few kcal/mol when examining the major trends in hydrocarbon chemistry, it appears that the difference in ΔE_a^* energy between a saturated C_nH_{2n+2} hydrocarbon and an unsaturated C_nH_{2n} hydrocarbon is roughly due to the removal of two CH bonds ($2 \times 106.806\,\mathrm{kcal/mol}$)

and their replacement by a CC bond (69.633 kcal/mol), representing a decrease of $\sim$144.0 kcal/mol in ΔE_a^* energy for each newly created double bond.

With ΔE_a^* in kcal/mol, a few examples are (experimental values are given in parentheses): propane/propene, 860.1 (858.56); butane/butene, 1154.2 (1154.5); pentane/pentene, 1448.2 (1447.6); cyclohexane/cyclohexene, 1616.8 (1616.1); methylcyclohexane/1-methylcyclohexene, 1913.1 (1913.2); cyclohexane/1,4-cyclo-hexadiene, 1472.8 (1472.8), and butane/butadiene, 1010.2 (1010.1). (Where appropriate, the averages of the various alkene isomers were used.)

The carbon–carbon triple bond in *acetylene* is not far from being the triple of a single-bond contribution; the factor suggested by Hartree–Fock calculations is $\sim$2.9. Thus, subtracting four CH contributions from butane and adding 1.9 $\times$ 69.633 kcal/mol (i.e., by subtracting $\sim$295 kcal/mol from the ΔE_a^* of butane) we obtain, in this rough estimate, ΔE_a^* (2-butyne) $\simeq$ 1003.1 kcal/mol. Under these circumstances, the agreement with the experimental value, $\Delta E_a^* = 1001.5$ kcal/mol is certainly satisfactory. Similarly, using now the 1.6 factor deduced for *benzene*, we estimate its atomization energy from that of cyclohexane by removing six CH bonds and adding 6 times 0.6 $\times$ 69.633 kcal/mol. This estimate leads to ΔE_a^*(benzene) $\simeq$ 1370.6 kcal/mol, still in rough agreement with its experimental value, namely, $\Delta E_a^* = 1366.5$ kcal/mol. Finally, taking propane as a precursor for allene, $H_2C{=}C{=}CH_2$, we replace four CH bonds by two CC bonds using the 128.5/66.7 ratio of carbon–carbon bond energies indicated by HF calculations. The value thus deduced for allene, $\Delta E_a^* \simeq 705.9$ kcal/mol, is reasonably close to its experimental counterpart, 701.9 kcal/mol.

No claim for accuracy can be made for brute-force approaches of this sort. It remains, however, that a most instructive link is created in this manner between typical saturated, olefinic, acetylenic, and aromatic hydrocarbons. This is rewarding; all the pieces seem to fall in place in a very orderly fashion.

The atomization energies of *dialkylethers* are conveniently deduced from the following general formula (in kcal/mol units):

$$\Delta E_a^* = \sum_{k<l} \varepsilon_{kl}^\circ + 0.0356 \sum N_{CC}\delta_C + 0.0529 \sum \delta_{C \neq \alpha} + 0.1217 \sum \delta_{C_\alpha}$$

$$+ A_4\delta_O + 7.393n + 0.92 \sum N_{C_\alpha C} - 10.67 - E_{nb} \qquad \text{(A.22)}$$

The standard bond contributions ε_{kl}° are 69.633 for *any* CC, 106.806 for *any* CH and 79.78 kcal/mol for one CO bond. The $\delta_{C \neq \alpha}$ shifts are relative to ethane (at 5.8 ppm from TMS), and the δ_{C_α} shifts are relative to the α-carbon of diethylether (65.9 ppm from TMS). The $\sum N_{CC}\,\delta_C$ term includes both δ_{C_α} and $\delta_{C \neq \alpha}$. The ^{17}O NMR shifts are relative to that of diethylether (6.5 ppm from water). The parameter n is the number of C atoms and $\sum N_{C_\alpha C}$ is the number of CC bonds formed by the two α-carbons. One uses $A_4 = 0.1007$ when $\delta_O < 0$ and $A_4 = 0.0994$ kcal mol^{-1} ppm^{-1} when $\delta_O > 0$. Rough evaluations of the nonbonded part, E_{nb}, can be made as follows. For normal alkyl groups larger than C_2H_5, E_{nb} becomes more negative by $\sim$0.13 kcal/mol for each added CH_2 group, but branching (e.g., for *iso*-C_3H_7

instead of n-C_3H_7) makes E_{nb} more positive by ~ 0.08 kcal/mol. Acetals should not be calculated in this manner because their charge-^{17}O NMR shift relationship differs from that of the simple ethers.

The energy formula describing noncyclic carbonyl compounds is (in kcal/mol units):

$$\Delta E_a^* = \sum_{k<l} \varepsilon_{kl}^\circ + 0.0356 \sum N_{CC}\delta_C + 0.0529 \sum \delta_{C\neq\alpha} + 0.0675\delta_{C_\alpha}$$

$$+ A_4'\delta_O + 7.393n + 5.07 N_{C_aC} - 19.42 - E_{nb} \tag{A.23}$$

$$A_4' \simeq -(0.098 + 1.16 \times 10^{-3} \times \delta_O) \text{ kcal mol}^{-1} \text{ ppm}^{-1} \tag{A.24}$$

where n is the number of carbon atoms. N_{C_aC} is the number of CC bonds formed by the C atom attached to oxygen. $N_{CC}\delta_C$ includes both δ_{C_α} (from the acetone carbonyl-C atom) and $\delta_{C\neq\alpha}$ (from ethane) for the atoms not bonded to O. The appropriate ε_{kl}° bond parameters are, in kcal/mol units, $\varepsilon_{CC}^\circ = 69.633$, $\varepsilon_{CH}^\circ = 106.806$ (as in ethane), and $\varepsilon_{CO}^\circ = 179.40$ (like the CO bond in acetone).

[To gain familiarity with Eqs. (A.15)–(A.18), (A.22), and (A.23), the input data ($\sum N_{CC}\delta_C$, $\sum \delta_C$, etc.) listed in Refs. 44, 27, and 141 are most useful. The ^{13}C NMR shifts of the dienes are reported in Ref. 192.]

A.3 BOND ENERGY FORMULAS

Any bond energy formula can be expressed either i) by reference to a selected bond with reference net atomic charges q_k° and q_l° at the bond-forming atoms k and l, or ii) by reference to hypothetical k–l bonds constructed with the assumption $q_k^\circ = q_l^\circ = 0$. The former reflects a physical situation, but requires additional work in order to satify charge normalization constraints; it is most useful in the construction of general energy formulas for molecules that use chemical shifts espressed with respect to the appropriate references. The latter method simplifies bond-by-bond calculations. The two forms are

$$\varepsilon_{kl} = \varepsilon_{kl}^\circ + a_{kl}(q_k - q_k^\circ) + a_{lk}(q_l - q_l^\circ) \tag{A.25}$$
$$= \varepsilon_{kl}^{\circ\prime} + a_{kl}q_k + a_{lk}q_l \tag{A.26}$$

with

$$\varepsilon_{kl}^{\circ\prime} = \varepsilon_{kl}^\circ - a_{kl}q_k^\circ - a_{lk}q_l^\circ \tag{A.27}$$

Where appropriate, ε_{kl}° and $\varepsilon_{kl}^{\circ\prime}$ are corrected to include effects arising from possible shifts of π orbitals and conjugation.

In the following, all energies are expressed in kcal/mol and the charges in me (10^{-3} electron) units.

- $C(sp^3)$—$C(sp^3)$ *Carbon–Carbon Bonds* (alkanes)

$$\varepsilon_{CC} = 69.633 - 0.488(q_{C_k} - 35.1) - 0.488(q_{C_l} - 35.1) \tag{A.28}$$
$$= 103.891 - 0.488q_{C_k} - 0.488q_{C_l} \tag{A.29}$$

- $C(sp^3)$—H *Carbon–Hydrogen Bonds* (alkanes)

$$\varepsilon_{CH} = 106.806 - 0.247(q_{C_k} - 35.1) - 0.632(q_{H_l} + 11.7) \quad \text{(A.30)}$$
$$= 108.081 - 0.247q_{C_k} - 0.632q_{H_l} \quad \text{(A.31)}$$

- $C(sp^3)$—N *Carbon–Nitrogen Bonds* (amines)

$$\varepsilon_{CN} = 60.44 - 0.603(q_{C_k} - 31.77) - 0.448(q_{N_l} + 9.00) \quad \text{(A.32)}$$
$$= 75.56 - 0.603q_{C_k} - 0.448q_{N_l} \quad \text{(A.33)}$$

- $C(sp^3)$—O *Carbon–Oxygen Bonds* (ethers)

$$\varepsilon_{CO} = 79.78 - 0.712(q_{C_k} - 31.26) - 0.501(q_{O_l} - 5.18) \quad \text{(A.34)}$$
$$= 104.635 - 0.712q_{C_k} - 0.501q_{O_l} \quad \text{(A.35)}$$

- $C(sp^3)$—O *Carbon–Oxygen Bonds* (alcohols)

$$\varepsilon_{CO} = 104.635 - 0.712q_{C_k} - 0.649q_{O_l} \quad \text{(A.36)}$$

- N—H *Nitrogen–Hydrogen Bonds* (amines)

$$\varepsilon_{NH} = 101.36 - 0.197q_{N_k} - 0.794q_{H_l} \quad \text{(A.37)}$$

- O—H *Oxygen–Hydrogen Bonds* (alcohols)

$$\varepsilon_{OH} = 115.678 - 0.400q_{O_k} - 1.000q_{H_l} \quad \text{(A.38)}$$

Carbonyl carbon atoms form the following bonds:

- $C{=}O$ *Carbon–Oxygen Double Bonds* (acetone)

$$\varepsilon_{CO} = 179.40 - 0.742(q_C - 14.0)$$
$$- [0.668_3 + 1.59 \times 10^{-4}(q_O + 21.2)](q_O + 21.2) \quad \text{(A.39)}$$
$$= 175.55 - 0.742q_C - 0.675q_O - 1.59 \times 10^{-4}q_O^2 \quad \text{(A.40)}$$

- $C({=}O)$—$C(sp^3)$ *Carbonyl-C Bonded to* sp^3 *Carbon* (acetone)

$$\varepsilon_{CC} = 69.633 - 0.413(q_{C_k} - 35.1) - 0.488(q_{C_l} - 35.1) \quad \text{(A.41)}$$
$$= 101.26 - 0.413q_{C_k} - 0.488q_{C_l} \quad \text{(A.42)}$$

- $C({=}O)$—H *Carbonyl-C Bonded to Hydrogen* (aldehydes)

$$\varepsilon_{CH} = 106.806 - 0.173(q_{C_k} - 35.1) - 0.632(q_{H_l} + 11.7) \quad \text{(A.43)}$$
$$= 105.48 - 0.173q_{C_k} - 0.632q_{H_l} \quad \text{(A.44)}$$

The following bonds occur in olefins:

- $C(sp^2)\!=\!C(sp^2)$ *Carbon–Carbon Double Bonds* (ethylene)

$$\varepsilon_{C=C} = 139.37 - 0.183(q_{C_k} - 7.7) - 0.183(q_{C_l} - 7.7) \tag{A.45}$$
$$= 142.19 - 0.183q_{C_k} - 0.183q_{C_l} \tag{A.46}$$

- $C(sp^2)\!-\!C(sp^2)$ *Carbon–Carbon Single Bonds* (butadiene)^2

$$\varepsilon_{CC} = 89.14 + 0.258(q_{C_k} - 7.7) + 0.258(q_{C_l} - 7.7) \tag{A.47}$$
$$= 85.17 + 0.258q_{C_k} + 0.258q_{C_l} \tag{A.48}$$

- $C(sp^3)\!-\!C(sp^2)$ *Carbon–Carbon Bonds* (olefins)^3

$$\varepsilon_{CC} = 77.67 - 0.488(q_{C_k} - 35.1) + 0.275(q_{C_l} - 7.7) \tag{A.49}$$
$$= 92.68 - 0.488q_{C_k} + 0.275q_{C_l} \tag{A.50}$$

- $C(sp^2)\!-\!H$ *Carbon–Hydrogen Bonds* (olefins)

$$\varepsilon_{CH} = 110.69 + 0.454(q_{C_k} - 7.7) - 0.632(q_{H_l} + 11.7) \tag{A.51}$$
$$= 99.80 + 0.454q_{C_k} - 0.632q_{H_l} \tag{A.52}$$

The following bonds occur in benzenoid hydrocarbons. Consider first the endocyclic carbon–carbon bonds, namely, those found in benzene, with $\varepsilon_{CC}^\circ = 115.39$ (or $\varepsilon_{CC}^\circ = 124.84$) kcal/mol, which—in a sketchy way—are some kind of averages between a single and a double sp^2–sp^2 bond. (Their number is double that of the number of double bonds that can be written in classical Kekulé structures, e.g., 2×5 in naphthalene, 2×7 in anthracene.) But in polynuclear benzenoid structures there are not twice as many "averages" as there are Kekulé double bonds. Hence, consider the "extra" single $C(sp^2)\!-\!C(sp^2)$ bonds like the one found in naphthalene, or the two "extra" single bonds found in anthracene. The appropriate bond energy formulas are

- $C(Ar)\!:\!:\!:C(Ar)$ *(Endo) Carbon–Carbon Bonds* (benzene)

$$\varepsilon_{CC} = 115.39 - 0.358(q_{C_k} - 13.2) - 0.358(q_{C_l} - 13.2) \tag{A.53}$$
$$= 124.84 - 0.358q_{C_k} - 0.358q_{C_l} \tag{A.54}$$

- $C(Ar)\!-\!C(Ar)$ *(Endo) Carbon–Carbon "Single" Bonds*

$$\varepsilon_{CC} = 91.21 - 0.358(q_{C_k} - 13.2) - 0.358(q_{C_l} - 13.2) \tag{A.55}$$
$$= 100.66 - 0.358q_{C_k} - 0.358q_{C_l} \tag{A.56}$$

$^2 a_{CC}^{\sigma\pi} = 0.258$ follows from $a_{CC}^\sigma(1.463) = -0.467_5$, $a_{CC}^\pi(1.485) = -0.4348$ kcal mol^{-1} me^{-1} and $m = -0.955$.
$^3 a_{C'C}^{\sigma\pi} = 0.275$ follows from $a_{C'C}^\sigma(1.53) = -0.450_{55}$, $a_{C'C}^\pi(1.55) = -0.417_{92}$ kcal mol^{-1} me^{-1} and $m = -0.955$.

In Eqs. (A.53) and (A.55), for C(Ar) carbons engaged in exocyclic CC bonds, use $a_{CC}^{\sigma\pi}(1.40) = -0.249$, $m = -0.9$, instead of -0.358 ($m = -0.814$) kcal mol^{-1} me^{-1}. The carbon–hydrogen bonds formed by the arylcarbon atoms are

- C(Ar)—H *Carbon–Hydrogen Bonds* (aromatic molecules)

$$\varepsilon_{CH} = 111.41 - 0.083(q_{C_k} - 13.2) - 0.632(q_{H_l} + 11.7) \tag{A.57}$$
$$= 105.11 - 0.083q_{C_k} - 0.632q_{H_l} \tag{A.58}$$

All the a_{kl} parameters indicated in Eqs. (A.53)–(A.58) were calculated by means of the general formula

$$a_{kl}^{\sigma\pi} = \frac{m a_{kl}^{\sigma} + a_{kl}^{\pi}}{1 + m} \tag{A.59}$$

with $m = -0.814$. When the ring carbons are engaged in *exocyclic* carbon–carbon bonds, as in toluene, styrene, or biphenyl, for example, one must use $a_{CC}^{\sigma}(1.48) = -0.463_1$ and $a_{CC}^{\pi}(1.48) = -0.439_4$ kcal mol^{-1} me^{-1}, with $m = -0.9$. The appropriate formulas are

- C(Ar)—C(sp^3) *Exocyclic Carbon–Carbon Bonds* (toluene)

$$\varepsilon_{CC} = 79.33 - 0.226(q_{C_k} - 13.2) - 0.488(q_{C_l} - 35.1) \tag{A.60}$$
$$= 99.44 - 0.226q_{C_k} - 0.488q_{C_l} \tag{A.61}$$

- C(Ar)—C(Ar) *Exocyclic Carbon–Carbon Bond* (biphenyl)

$$\varepsilon_{CC} = 88.89 - 0.226(q_{C_k} - 13.2) - 0.226(q_{C_l} - 13.2) \tag{A.62}$$
$$= 94.86 - 0.226q_{C_k} - 0.226q_{C_l} \tag{A.63}$$

- C(Ar)—C(sp^2) *Exocyclic C—C Bonds* (styrene)

$$\varepsilon_{CC} = 89.69 - 0.226(q_{C_k} - 13.2) + 0.275(q_{C_l} - 7.7) \tag{A.64}$$
$$= 90.56 - 0.226q_{C_k} + 0.275q_{C_l} \tag{A.65}$$

Similar formulas apply to the *cis-* and *trans-*stilbenes:

- C(Ar)—C(sp^2) *Exocyclic C—C Bonds* (cis-stilbene)

$$\varepsilon_{CC} = 86.77 - 0.226(q_{C_k} - 13.2) + 0.275(q_{C_l} - 7.7) \tag{A.66}$$
$$= 87.64 - 0.226q_{C_k} + 0.275q_{C_l} \tag{A.67}$$

- $C(Ar)$—$C(sp^2)$ *Exocyclic C—C Bonds* (*trans*-stilbene)

$$\varepsilon_{CC} = 90.14 - 0.226(q_{C_k} - 13.2) + 0.275(q_{C_l} - 7.7) \qquad (A.68)$$
$$= 91.01 - 0.226q_{C_k} + 0.275q_{C_l} \qquad (A.69)$$

Not unexpectedly, numerous different types of carbon–carbon bonds should be considered in aromatic molecules, depending on the types of atoms involved (aromatic sp^2 carbons, olefinic sp^2 carbons and sp^3 carbon atoms) and the possible occurrence of conjugation, as well as geometric features that possibly counteract conjugation. It is thus clear that—contrasting with straightforward quantum chemistry—some advance knowledge is necessary in our calculations, for example, about geometry and the appropriate atomic charges, which set the limits of our methods.

But it is also true that the present use of transferable bond energy parameters, such as those found in Eqs. (A.47)–(A.69), translate in a conceptually simple way the logics associated with shifts of π-electron centroids, with bond lengths, and with the appropriate charge redistributions. The results surely illustrate the flexibility of the method capable of generating the parameters adapted to the various situations that are encountered. Of course, this demands thought.

In turn, it also appears that "chemical insight" has been gained about the various contributions to molecular stability, which can now be dealt with efficiently. In a way, chemistry has been given back to chemists who look for general rules that may guide them into the future. Considering the conceptual simplicity of the methods described here, the quality of the tests involving experimental data is certainly encouraging. No doubt, the proposed parameters could be improved in the future, but it also seems wise to wait until forthcoming technical refinements improve on the quality of the experimental results to be used in comparisons with theory.

BIBLIOGRAPHY

1. T. L. Allen, *J. Chem. Phys.* **31**:1039 (1959).

2. K. Pihlaja, K. Rossi, and P. Valniotalo, *J. Chem. Eng. Data* **30**:387 (1985).

3. T. Berlin, *J. Chem. Phys.* **19**:208 (1951).

4. K. Ruedenberg, in *Localization and Delocalization in Quantum Chemistry*, Reidel, Dordrecht, Netherlands, 1975, Vol. 1, 223.

5. W. Kutzelnigg, in *The Concept of the Chemical Bond*, Z. B. Maksić (ed.), Springer-Verlag, Berlin, 1990, Part 2, p. 1.

6. R. F. W. Bader, *Atoms in Molecules, a Quantum Theory*, Clarendon Press, Oxford, 1990.

7. L. H. Thomas, *Proc. Cambridge Philos. Soc.* **23**:542 (1926).

8. E. Fermi, *Z. Phys.* **48**:73 (1928).

9. P. Politzer and R. G. Parr, *J. Chem. Phys.* **61**:4258 (1974).

10. J. A. Alonso, D. J. González, and L. C. Balbás, *Int. J. Quantum Chem.* **22**:989 (1982).

11. N. H. March, *Self-Consistent Fields in Atoms*, Pergamon, Oxford, 1975.

12. N. H. March, in *Theoretical Chemistry, a Specialist Periodical Report*, The Royal Society of Chemistry, 1981, Vol. 4, p. 92.

13. V. Barone and S. Fliszár, *Int. J. Quantum Chem.* **55**:469 (1995).

14. V. Barone and S. Fliszár, *J. Mol. Struct.* (*Theochem*) **369**:29 (1996).

15. W. J. Hehre, R. Ditchfield, and J. A. Pople, *J. Chem. Phys.* **56**:2257 (1972).

16. R. Krishnan, J. S. Binkley, R. Seeger, and J. A. Pople, *J. Chem. Phys.* **72**:650 (1980).

17. M. J. Frisch, J. A. Pople, and J. S. Binkley, *J. Chem. Phys.* **80**:3265 (1984).

Atomic Charges, Bond Properties, and Molecular Energies, by Sándor Fliszár
Copyright © 2009 John Wiley & Sons, Inc.

18. F. B. van Duijneveldt, Technical Report RJ 945, IBM Research Laboratories, San Jose, CA, 1971.

19. A. D. Becke, *Phys. Rev. B* **38**:3098 (1988).

20. C. Lee, W. Yang, and R. G. Parr, *Phys. Rev. B* **37**:785 (1988).

21. V. H. Smith, Jr., *Phys. Scripta* **15**:147 (1977); A. Julg, *Top. Curr. Chem.* **58**:1 (1975).

22. P. Politzer, K. C. Leung, J. D. Elliott, and S. K. Peters, *Theor. Chim. Acta* **38**:101 (1975).

23. R. F. W. Bader, *MTP Int. Rev. Sci.; Phys. Chem. Ser. 2* **1**:43 (1975); *Acc. Chem. Res.* **8**:34 (1975).

24. J. R. Van Wazer and I. Abser, *Electron Densities in Molecules and Molecular Orbitals*, Academic Press, New York, 1975.

25. R. F. W. Bader, S. G. Anderson, and A. J. Duke, *J. Am. Chem. Soc.* **101**:1389 (1979).

26. K. B. Wiberg, *J. Am. Chem. Soc.* **102**:1229 (1980).

27. S. Fliszár, *Charge Distributions and Chemical Effects*, Springer-Verlag, Berlin/ Heidelberg, 1983.

28. C. F. Guerra, J.-W. Handgraaf, E. J. Baerends, and F. M. Bickelhaupt, *J. Comput. Chem.* **25**:189 (2004).

29. G. Del Re, *J. Chem. Soc.* 4031 (1958).

30. G. Del Re, S. Fliszár, M. Comeau, and C. Mijoule, *Can. J. Chem.* **63**:1487 (1985).

31. R. S. Mulliken, *J. Chem. Phys.* **23**:1833 (1955); see also pp. 1841, 2338, 2343.

32. P.-O. Löwdin, *J. Chem. Phys.* **21**:374 (1953).

33. K. Jug, *Theor. Chim. Acta* **23**:183 (1971); *ibid.* **26**:231 (1972); *ibid.* **29**:9 (1973).

34. S. Fliszár and S. Chrétien, *J. Mol. Struct. (Theochem)* **618**:133 (2002).

35. S. Fliszár and S. Chrétien, *J. Mol. Struct. (Theochem)* **668**:217 (2004).

36. S. Fliszár, S. Chrétien, and E. C. Vauthier, *J. Mol. Struct. (Theochem)* **685**:175 (2004).

37. S. Fliszár and S. Chrétien, *J. Mol. Struct. (Theochem)* **668**:101 (2004).

38. S. Fliszár, A. Goursot, and H. Dugas, *J. Am. Chem. Soc.* **96**:4358 (1974).

39. G. Kean, D. Gravel, and S. Fliszár, *J. Am. Chem. Soc.* **98**:4749 (1976).

40. H. Henry and S. Fliszár, *J. Am. Chem. Soc.* **100**:3312 (1978).

41. M.-T. Béraldin, E. C. Vauthier, and S. Fliszár, *Can. J. Chem.* **60**:106 (1982).

42. S. Fliszár, G. Cardinal, and M.-T. Béraldin, *J. Am. Chem. Soc.* **104**:5287 (1982).

43. M. Comeau, M.-T. Béraldin, E. Vauthier, and S. Fliszár, *Can. J. Chem.* **63**:3226 (1985).

44. S. Fliszár, *Atoms, Chemical Bonds and Bond Dissociation Energies*, Lecture Notes in Chemistry, Vol. 63, Springer-Verlag, Berlin/Heidelberg, 1994.

45. I. N. Levine, *Quantum Chemistry*, 3rd ed., Allyn & Bacon, Newton, MA, 1983.

46. H. Eyring, J. Walter, and G. E. Kimball, *Quantum Chemistry*, Wiley, New York, 1954.

47. D. R. Hartree, *The Calculation of Atomic Structures*, Wiley, New York, 1957; C. Froese Fischer, *The Hartree–Fock Method for Atoms: A Numerical Approach*, Wiley, New York, 1977.

48. J. B. Mann, *Atomic Structure Calculations*, Los Alamos Scientific Laboratory, Univ. California, Los Alamos, NM, Part I: *Hartree–Fock Energy Results for the Elements Hydrogen to Lawrencium*, 1967; Part II: *Hartree–Fock Wave Functions and Radial Expectation Values*, 1968.

49. R. G. Parr, *Quantum Theory of Molecular Electronic Structure*, Benjamin, New York, 1963.

50. F. Grimaldi, A. Lecourt, and C. Moser, *Int. J. Quantum Chem.* **51**:153 (1967).

51. C. Mijoule, J.-M. Leclercq, M. Comeau, S. Fliszár, and M. Picard, *Can. J. Chem.* **70**:68 (1992); C. Mijoule, J.-M. Leclercq, S. Odiot, and S. Fliszár, *Can. J. Chem.* **63**:1741 (1985).

52. R. N. Dixon and I. L. Robertson, *Specialist Periodical Reports, Theoretical Chemistry*, Vol. 3, The Chemical Society, London, 1978, Chap. 4.

53. G. Delgado-Barrio and R. F. Prat, *Phys. Rev. A* **12**:2288 (1975).

54. P. Sjoberg, J. S. Murray, T. Brinck, and P. Politzer, *Can. J. Chem.* **68**:1440 (1990).

55. P. Politzer, J. S. Murray, and M. E. Grice, *Charge Capacities and Shell Structures of Atoms, Structure and Bonding*, Springer-Verlag, Berlin/Heidelberg, 1993, Vol. 80, p. 101.

56. P. Politzer, J. S. Murray, M. E. Grice, T. Brinck, and Ranganathan, *J. Chem. Phys.* **95**:6699 (1991).

57. P. Politzer, *J. Chem. Phys.* **72**:3027 (1980).

58. K. D. Sen, T. V. Gayatri, R. Krishnaveni, M. Kakkar, H. Toufar, G. O. A. Janssens, B. G. Baekelandt, R. A. Schoonheydt, and W. J. Mortier, *Int. J. Quantum Chem.* **56**:399 (1995).

59. W.-P. Wang and R. G. Parr, *Phys. Rev. A*, **16**:891 (1977).

60. H. Weinstein, P. Politzer, and S. Srebrenik, *Theor. Chim. Acta* **38**:159 (1975).

61. P. Politzer and R. G. Parr, *J. Chem. Phys.* **64**:4634 (1976).

62. J. Goodisman, *Phys. Rev. A* **2**:1193 (1970); *Theor. Chim. Acta* **24**:1 (1972).

63. P. Gombás, *Die Statistische Theorie des Atoms und ihre Anwendungen*, Springer-Verlag, Vienna, 1949.

64. E. Clementi, *Tables of Atomic Functions*, IBM Corp., San Jose, CA, 1965.

65. H. Schmider, R. P. Sagar, and V. H. Smith, Jr., *Can. J. Chem.* **70**:506 (1992).

66. A. M. Simas, R. P. Sagar, A. C. T. Ku, and V. H. Smith, Jr., *Can. J. Chem.* **66**:1923 (1988).

67. H. Schmider, R. P. Sagar, and V. H. Smith, Jr., *J. Chem. Phys.* **94**:8627 (1991).

68. S. Fliszár, N. Desmarais, and G. Dancausse, *Can. J. Chem.* **70**:537 (1992); N. Desmarais, PhD thesis, Univ. Montreal, Montreal, 1995.

69. N. Desmarais and S. Fliszár, *Theor. Chim. Acta* **94**:187 (1996).

70. E. Clementi and C. Roetti, *At. Data Nucl. Data Tables* **14**:179 (1974).

71. P. Politzer and K. C. Daiker, *Chem. Phys. Lett.* **20**:309 (1973).

72. P.-O Widmark, P.-Å. Malmqvist, and B. O. Roos, *Theor. Chim. Acta* **77**:291 (1990); P.-O Widmark, B. J. Persson, and B. O. Roos, *Theor. Chim. Acta* **79**:419 (1991).

73. N. Desmarais, G. Dancausse, and S. Fliszár, *Can. J. Chem.* **71**:175 (1993).

74. H. Hellmann, *Einführung in die Quantum Chemie*, Deutike, Leipzig, 1937; R. P. Feynman, *Phys. Rev.* **56**:340 (1939).

75. E. A. Milnes, *Proc. Cambridge Philos. Soc.* **23**:794 (1927).

76. N. H. March, *Adv. Phys.* **6**:1 (1957).

77. R. G. Parr, S. R. Gadre, and L. J. Bartolotti, *Proc. Natl. Acad. Sci. USA* **76**:2522 (1979).

78. K. Ruedenberg, *J. Chem. Phys.* **66**:375 (1977).

79. P. Politzer, *J. Chem. Phys.* **64**:4239 (1976); P. Politzer, *ibid.* **69**:491 (1978).

80. C. E. Moore, Natl. Stand. Ref. Data Ser., US Natl. Bureau of Standards, 1970, p. 34.

81. S. N. Datta, *Int. J. Quantum Chem.* **56**:91 (1995).

82. J. P. Desclaux, *At. Data Nucl. Data Tables* **12**:311 (1973).

83. S. Fliszár, *Theor. Chem. Acc.* **96**:122 (1997).

84. L. R. Kahn, P. Baybutt, and D. G. Truhlar, *J. Chem. Phys.* **65**:3826 (1976).

85. T. H. Dunning *J. Chem. Phys.* **53**:2823 (1970).

86. K. Pierloot, B. Dumez, P.-O. Widmark, and B. O. Roos, *Theor. Chim. Acta* **90**:87 (1995).

87. C. Froese Fischer, *Comput. Phys. Commun.* **1**:151 (1972); **4**:107 (1972); **7**:236 (1972); *J. Comput. Phys.* **10**:211 (1972).

88. G. Kean, MSc thesis, Univ. Montreal, Montreal, 1974.

89. M. E. Schwartz, *Chem. Phys. Lett.* **6**:631 (1970).

90. S. Fliszár, M. Foucrault, M.-T. Béraldin, and J. Bridet, *Can. J. Chem.* **59**:1074 (1981).

91. R. G. Parr, P. W. Ayers, and R. F. Nalewajski, *J. Phys. Chem. A* **109**:3957 (2005).

92. E. A. Laws and W. N. Lipscomb, *Israel J. Chem.* **10**:77 (1972).

93. P. E. Cade, R. F. W. Bader, W. H. Hennecker, and I. Keaveny, *J. Chem. Phys.* **50**:5313 (1969); R. F. W. Bader and A. D. Bandrauk, *J. Chem. Phys.* **49**:1653 (1968); R. F. W. Bader, W. H. Hennecker, and P. Cade, *J. Chem. Phys.* **46**:3341 (1967); R. F. W. Bader and W. H. Hennecker, *J. Am. Chem. Soc.* **88**:280 (1966); *ibid.* **87**:3063 (1965).

94. P. Coppens, in *International Review of Science*, Physical Chemistry, Series 2, Vol. 2, J. M. Robertson (ed.), Butterworths, London, 1975, pp. 21–56; P. Coppens and E. D. Stevens, *Adv. Quantum Chem.* **10**:1 (1977); B. Dawson, *Studies of Atomic Charge Density by X-Ray and Neutron Diffraction. A Perspective*, Pergamon Press, London, 1975; F. L. Hirschfeld and S. Rzotkiewicz, *Mol. Phys.* **27**:319 (1974); B. J. Ransil and J. J. Sinai, *J. Am. Chem. Soc.* **94**:7268 (1972); B. J. Ransil and J. J. Sinai, *J. Chem. Phys.* **46**:4050 (1967).

95. M. Roux, M. Cornille, and L. Burnelle, *J. Chem. Phys.* **37**:933 (1972); M. Roux, *J. Chim. Phys.* **57**:53 (1960); *ibid.* **55**:754 (1958); M. Roux, S. Besnainou, and R. Daudel, *J. Chim. Phys.* **53**:218, 939 (1956).

96. S. Fliszár, G. Kean, and R. Macaulay, *J. Am. Chem. Soc.* **96**:4353 (1974).

97. R. W. Taft, in *Steric Effects in Organic Chemistry*, M. S. Newman (ed.), Wiley, New York, 1956; R. W. Taft, *J. Am. Chem. Soc.* **75**:4231 (1953).

98. S. Fliszár, *Can. J. Chem.* **54**:2839 (1976).

99. D. R. Salahub and C. Sándorfy, *Theor. Chim. Acta* **20**:227 (1971).

100. S. Diner, J.-P. Malrieu, and P. Claverie, *Theor. Chim. Acta* **13**:1 (1969); J.-P. Malrieu, P. Claverie, and S. Diner, *Theor. Chim. Acta* **13**:18 (1969); S. Diner, J.-P. Malrieu, F. Jordan, and M. Gilbert, *Theor. Chim. Acta* **13**:101 (1969); S. Fliszár and J. Sygusch, *Can. J. Chem.* **51**:991 (1973).

101. R. Hoffmann, *J. Chem. Phys.* **39**:1397 (1963); J. M. Sichel and M. A. Whitehead, *Theor. Chim. Acta* **5**:35 (1966).

102. G. Kean and S. Fliszár, *Can. J. Chem.* **52**:2772 (1974).

103. J. M. André, P. Degand, and G. Leroy, *Bull. Soc. Chim. Belg.* **80**:585 (1971).

104. A. E. Foti, V. H. Smith, and S. Fliszár, *J. Mol. Struct. (Theochem)* **68**:227 (1980).

105. V. W. Laurie and J. S. Muenter, *J. Am. Chem. Soc.* **88**:2883 (1966).

106. E. C. Vauthier and S. Fliszár, manuscript in preparation.

107. R. Roberge and S. Fliszár, *Can. J. Chem.* **53**:2400 (1975).

108. S. Fliszár, *J. Am. Chem. Soc.* **102**:6946 (1980).

109. M.-T. Béraldin and S. Fliszár, *Can. J. Chem.* **61**:197 (1983).

110. R. S. Mulliken and C. C. Roothaan, *Chem. Rev.* **41**:219 (1947).

111. K. B. Wiberg and J. J. Wendoloski, *J. Comput. Chem.* **2**:53 (1981).

112. M. S. Gordon and W. England, *J. Am. Chem. Soc.* **94**:5168 (1972).

113. R. H. Pritchard and C. W. Kern, *J. Am. Chem. Soc.* **91**:1631 (1969).

114. W. H. Flygare, *Molecular Structure and Dynamics*, Prentice-Hall, Englewood Cliffs, NJ, 1978.

115. E. C. Vauthier, S. Odiot, and F. Tonnard, *Can. J. Chem.* **60**:957 (1982).

116. J. A. Pople, J. W. McIver, and N. S. Ostlund, *J. Chem. Phys.* **49**:2960 (1968).

117. J. A. Pople and D. L. Beveridge, *Approximate Molecular Orbital Theory*, McGraw-Hill, New York, 1970.

118. R. Ditchfield, *Mol. Phys.* **27**:789 (1974).

119. E. Vauthier, S. Fliszár, F. Tonnard, and S. Odiot, *Can. J. Chem.* **61**:1417 (1983).

120. A. A. Germer, *Theor. Chim. Acta* **34**:245 (1974).

121. M. Jallah-Heravi and G. A. Webb, *Org. Magn. Reson.* **13**:116 (1980).

122. H. Spiesecke and W. G. Schneider, *Tetrahedron Lett.* 468 (1961).

123. G. A. Olah and G. D. Mateescu, *J. Am. Chem. Soc.* **92**:1430 (1970).

124. G. A. Olah, J. M. Bollinger, and A. M. White, *J. Am. Chem. Soc.* **91**:3667 (1969).

125. G. J. Ray, A. K. Colter, and R. J. Kurland, *Chem. Phys. Lett.* **2**:324 (1968).

126. E. A. La Lancette and R. E. Benson, *J. Am. Chem. Soc.* **87**:1491 (1965).

127. W. J. Hehre, R. W. Taft, and R. D. Topsom, *Prog. Phys. Org. Chem.* **12**:159 (1976).

128. P. C. Lauterbur, *J. Am. Chem. Soc.* **83**:1838 (1961).

129. S. Fliszár, G. Cardinal, and N. A. Baykara, *Can. J. Chem.* **64**:404 (1986).

130. J. Bridet, M.-T. Béraldin, and S. Fliszár, *Can. J. Chem.* **63**:2468 (1985).

131. A. Veillard and G. Del Re, *Theor. Chim. Acta* **2**:55 (1964); G. Del Re, U. Esposito, and M. Carpentieri, *Theor. Chim. Acta* **6**:36 (1966).

132. K. B. Wiberg, *Tetrahedron* **24**:1083 (1968).

133. C. Trindle and O. Sinanoğlu, *J. Chem. Phys.* **49**:65 (1968).

134. J. E. Lennard-Jones and J. A. Pople, *Proc. Roy. Soc. (Lond.)*, **A220**:446 (1950); *ibid.* **A210**:190 (1951).

135. C. Edmiston and K. Ruedenberg, *Rev. Mod. Phys.* **35**:457 (1963); *J. Chem. Phys.* **43**:S97 (1965).

136. C. Trindle and O. Sinanoğlu, *J. Am. Chem. Soc.* **91**:853 (1969).

137. C. Juan and H. S. Gutowsky, *J. Chem. Phys.* **37**:2198 (1962).

138. J. B. Stothers, *Carbon-13 NMR Spectroscopy*, Academic Press, New York, 1972.

139. S. Fliszár, E. Vauthier, A. Cossé-Barbi, and E. L. Cavalieri, *J. Mol. Struct.* (*Theochem*) **531**:387 (2000).

140. C. Delseth and J.-P. Kintzinger, *Helv. Chim. Acta* **61**:1327 (1978).

141. S. Fliszár and M.-T. Béraldin, *Can. J. Chem.* **60**:792 (1982).

142. C. Delseth and J.-P. Kintzinger, *Helv. Chim. Acta* **59**:466 (1976); Erratum *Helv. Chim. Acta* **59**:1411 (1976).

143. G. Del Re and C. Barbier, *Croat. Chem. Acta* **57**:787 (1984).

144. W. Kolos, *Theor. Chim. Acta* **54**:187 (1980).

145. W. A. Sokalski, *J. Chem. Phys.* **77**:4529 (1982).

146. J. G. M. van Duijneveldt-Van de Rijt and F. B. van Duijneveldt, *J. Mol. Struct. (Theochem)* **89**:185 (1982).

147. K. Watanabe, T. Nakayama, and J. R. Mottl, *J. Quant. Spectr. Rad. Trans.* **2**:369 (1962).

148. A. B. Cornford, D. C. Frost, F. G. Herring, and C. A. McDowell, *Can. J. Chem.* **49**:1135 (1971).

149. R. O. Duthaler and J. D. Roberts, *J. Am. Chem. Soc.* **100**:3889 (1978).

150. M. Witanowski, L. Stefaniak, and H. Januszewski, in *Nitrogen NMR*, M. Witanowski and G. A. Wegg (eds.), Plenum Press, London, 1973.

151. W. M. Litchman, M. Alei, Jr., and A. E. Florin, *J. Am. Chem. Soc.* **91**:6574 (1969); M. Alei, Jr., A. E. Florin, and W. M. Litchman, *J. Am. Chem. Soc.* **92**:4828 (1970).

152. R. O. Duthaler and J. D. Roberts, *J. Magn. Reson.* **34**:129 (1979).

153. R. L. Lichter and J. D. Roberts, *J. Am. Chem. Soc.* **94**:2495 (1972).

154. R. O. Duthaler, K. L. Williamson, D. D. Giannini, W. H. Bearden, and J. D. Roberts, *J. Am. Chem. Soc.* **99**:8406 (1977).

155. M. Witanowski and L. Stefaniak, *J. Chem. Soc. B* 1061 (1967).

156. M. Witanowski, T. Urbanski, and L. Stefaniak, *J. Am. Chem. Soc.* **86**:2568 (1964).

157. M. Witanowski, *Tetrahedron* **23**:4299 (1967); E. F. Mooney and P. H. Winson, *Ann. Rep. NMR Spectrosc.* **2**:125 (1969).

158. B. Bak, L. Hansen-Nygaard, and J. Rastrup-Andersen, *J. Mol. Spectrosc.* **2**:361 (1958).

159. J. E. Del Bene, *J. Am. Chem. Soc.* **101**:6184 (1979).

160. W. J. Hehre, L. Radom, and J. A. Pople, *J. Am. Chem. Soc.* **94**:1496 (1972).

161. M. Witanowski, L. Stefaniak, H. Januszewski, and G. A. Webb, *Tetrahedron* **27**:3129 (1971).

162. L. Stefaniak, J. D. Roberts, M. Witanowski, and G. A. Webb, *Org. Magn. Res.* **22**:201 (1984).

163. R. O. Duthaler and J. D. Roberts, *J. Am. Chem. Soc.* **100**:4969 (1978).

164. J. D. Roberts, F. J. Weigert, J. I. Kroschwitz, and H. J. Reich, *J. Am. Chem. Soc.* **92**:1338 (1970).

165. H. Eggert and C. Djerassi, *J. Am. Chem. Soc.* **95**:3710 (1973).

166. L. P. Lindeman and J. Q. Adams, *Anal. Chem.* **43**:1245 (1971).

167. H. F. Widing and L. S. Levitt, *Tetrahedron* **30**:611 (1974).

168. A. Streitwieser, *J. Phys. Chem.* **66**:368 (1962).

169. D. M. Grant and E. G. Paul, *J. Am. Chem. Soc.* **86**:2984 (1964).

170. H. Henry and S. Fliszár, *Can. J. Chem.* **52**:3799 (1974).

171. J. Grignon and S. Fliszár, *Can. J. Chem.* **52**:2766 (1974).

172. I. Mayer, *Int. J. Quantum Chem.* **23**:341 (1983).

173. I. Mayer, *Chem. Phys. Lett.* **96**:270 (1983).

174. J. C. Slater, *Adv. Quantum Chem.* **6**:1 (1973); J. C. Slater, *Int. J. Quantum Chem. Symp.* **3**:727 (1970); J. C. Slater, in *Computational Methods in Band Theory*, P. M. Marcus, J. F. Janak, and A. R. Williams (eds.), Plenum, New York/London, 1971, p. 447; J. C. Slater, *The Self-Consistent Field for Molecules and Solids*, Vol. 4, McGraw-Hill, Kuala-Lumpur, 1974.

175. K. H. Johnson, *Adv. Quantum Chem.* **7**:143 (1973); K. H. Johnson, *Ann. Rev. Phys. Chem.* **26**:39 (1975); J. W. D. Connolly, in G. A. Segal (ed.), *Modern Theoretical Chemistry*, Vol. IV, Plenum, New York, 1977.

176. N. Rösch, V. H. Smith, Jr., and M. W. Whangbo, *J. Am. Chem. Soc.* **96**:5984 (1974); *Inorg. Chem.* **15**:1768 (1976).

177. D. R. Stull and G. C. Sinke, *Adv. Chem. Ser.* **18** (1956).

178. K. S. Pitzer and E. Catalano, *J. Am. Chem. Soc.* **78**:4844 (1956); T. L. Cottrell, *J. Chem. Soc.* 1448 (1948); K. S. Pitzer, *J. Chem. Rev.* **27**:39 (1940); K. S. Pitzer, *J. Chem. Phys.* **8**:711 (1940); K. S. Pitzer, *J. Chem. Phys.* **5**:473 (1937).

179. T. Ot, A. Popowicz, and T. Ishida, *J. Phys. Chem.* **90**:3080 (1986).

180. S. Fliszár and J.-L. Cantara, *Can. J. Chem.* **59**:1381 (1981).

181. S. Fliszár, E. C. Vauthier, and S. Chrétien, *J. Mol. Struct. (Theochem)* **682**:153 (2004).

182. A. Scott and L. Radom, *J. Phys. Chem.* **100**:16502 (1996).

183. P. J. Stephens, F. J. Devlin, C. F. Chabalowski, and M. J. Frisch, *J. Phys. Chem.* **98**:11623 (1994).

184. A. D. Becke, *J. Chem. Phys.* **98**:5648 (1993).

185. S. H. Vosko, L. Wilk, and M. Nusair, *Can. J. Phys.* **58**:1200 (1980).

186. M. J. Frisch, G. W. Trucks, H. B. Schlegel, P. M. W. Gill, B. G. Johnson, M. A. Robb, J. R. Cheeseman, T. Keith, G. A. Peterson, J. A. Montgomery, K. Raghavachari, M. A. Al-Laham, V. G. Zakrzewski, J. V. Ortiz, M. W. Wong, J. L. Andres, E. S. Replogle, R. Gomperts, R. L. Martin, D. J. Fox, J. S. Binkley, D. J. Defrees, J. Baker, J. J. P. Stewart, M. Head-Gordon, C. Gonzalez, and J. A. Pople, GAUSSIAN 94, revision B1, Gaussian, Pittsburgh, PA, 1953.

187. R. G. Snyder and J. H. Schachtschneider, *Spectrochim. Acta* **21**:169 (1965).

188. F. D. Rossini, K. S. Pitzer, R. L. Arnett, R. M. Braun, and G. C. Pimentel, *Selected Values of Physical and Thermodynamic Properties of Hydrocarbons and Related Compounds*, American Petroleum Institute, Carnegie Press, Pittsburgh, PA, 1953.

189. R. H. Boyd, S. N. Sanwal, S. Shary-Tehrany, and D. McNally, *J. Phys. Chem.* **75**:1264 (1971).

190. H. Henry, G. Kean, and S. Fliszár, *J. Am. Chem. Soc.* **99**:5889 (1977).

191. S. Fliszár, F. Poliquin, I. Bădilescu, and E. Vauthier, *Can. J. Chem.* **66**:300 (1988).

192. S. Fliszár and G. Cardinal, *Can. J. Chem.* **62**:2748 (1984).

193. G. Herzberg, *Molecular Spectra and Molecular Structure. II. Infrared and Raman Spectra of Polyatomic Molecules*, Van Nostrand-Reinhold, New York, 1968.

194. S. S. Mitra and H. J. Bernstein, *Can. J. Chem.* **37**:553 (1959).

195. A. Bree, R. A. Kydd, T. N. Misra, and V. V. B. Vilkos, *Spectrochim. Acta* **A27**:2315 (1971).

196. A. Marchand and J.-P. Quintard, *Spectrochim. Acta* **A36**:941 (1980).

197. P. Debye, *Ann. Phys.* **39**:789 (1921).

198. O. K. Rice, *Statistical Mechanics, Thermodynamics and Kinetics*, Freeman, San Francisco, 1967.

199. A. Peluso and S. Fliszár, *Can. J. Chem.* **66**:2631 (1988).

200. A. Snelson, *J. Chem. Phys.* **74**:537 (1970).

201. J. Pacansky and B. Schrader, *J. Chem. Phys.* **78**:1033 (1983).

202. B. Schrader, J. Pacansky, and U. Pfeiffer, *J. Chem. Phys.* **88**:4069 (1984).

203. M. W. Chase, Jr., C. A. Davies, J. R. Downey, Jr., D. J. Frurip, R. A. McDonald, and A. N. Syverud, *JANAF Thermochemical Tables*, Vol. 1, Natl. Stand. Ref. Data Ser, Natl. Bureau of Standards (Suppl. 1), Vol. 14 1985.

204. M. E. Jacox, *J. Phys. Chem. Ref. Data* **13**:945 (1984).

205. M. Dupuis and J. J. Wendoloski, *J. Chem. Phys.* **80**:5696 (1984).

206. G. Del Re, *Gazz. Chim. Ital.* **102**:929 (1972).

207. S. Fliszár, E. C. Vauthier, and V. Barone, *Adv. Quantum Chem.* **36**:27 (2000).

208. S. Fliszár and M.-T. Béraldin, *Can. J. Chem.* **57**:1772 (1979).

209. L. Lunazzi, D. Macciantelli, F. Bernardi, and K. U. Ingold, *J. Am. Chem. Soc.* **99**:4573 (1977).

210. W. J. Hehre, L. Radom, P. v. R. Schleyer, and J. A. Pople, *Ab initio Molecular Orbital Theory*, Wiley, New York, 1986.

211. L. S. Bartell, K. Kuschitsu, and R. J. DeNui, *J. Chem. Phys.* **35**:1211 (1961); D. E. Shaw, D. W. Lepard, and H. L. Welsh, *J. Chem. Phys.* **42**:3736 (1965); K. Kuschitsu, *J. Chem. Phys.* **44**:906 (1966).

212. C. Froese Fischer, *At. Data Nucl. Data Tables*, **4**:301 (1972); *ibid.* **12**:87 (1973).

213. P. Politzer, *J. Chem. Phys.* **70**:1067 (1979).

214. K. Schwarz, *Phys. Rev. B* **5**:2466 (1972).

215. D. A. Case, M. Cook, and M. Karplus, *J. Chem. Phys.* **73**:3294 (1980).

216. M. W. Chase, Jr., *NIST-JANAF Thermochemical Tables*, 4th ed.; *J. Phys. Chem. Ref. Data, Monograph 9*, 1998.

217. L. V. Gurvich, I. V. Veyts, and C. B. Alcock, *Thermodynamic Properties of Individual Substances*, 4th ed., Hemisphere Publishing, New York, 1989.

218. Z. B. Maksić and K. Rupnik, *Croat. Chem. Acta* **56**:461 (1983).

219. R. G. Parr, *Int. J. Quant. Chem.* **26**:687 (1984).

220. R. S. Mulliken, *J. Am. Chem. Soc.* **88**:1849 (1966).

221. G. Del Re, *Adv. Quantum Chem.* **8**:95 (1974).

222. S. Fliszár, G. Del Re, and M. Comeau, *Can. J. Chem.* **63**:3551 (1985).

223. A. R. H. Cole, G. M. Mohay, and G. A. Osborne, *Spectrochim. Acta* **A23**:909 (1967); D. J. Marais, N. Sheppard, and B. P. Stoicheff, *Tetrahedron* **17**:163 (1962); A. Almenningen, O. Bastiansen, and M. Traetteberg, *Acta Chem. Scand.* **12**:1221 (1958); V. Schomaker and L. Pauling, *J. Am. Chem. Soc.* **61**:1769 (1939).

224. W. J. Hehre and J. A. Pople, *J. Am. Chem. Soc.* **97**:6941 (1975); P. N. Skancke and J. E. Boggs, *J. Mol. Struct.* **16**:179 (1973); B. Dunbacher, *Theor. Chim. Acta* **23**:346 (1972); U. Pincelli, B. Caldoli, and B. Levy, *Chem. Phys. Lett.* **13**:249 (1972); L. Radom and J. A. Pople, *J. Am. Chem. Soc.* **92**:4786 (1970).

225. K. Kuchitsu, T. Fukuyama, and Y. Morino, *J. Mol. Struct.* **1**:463 (1967); W. Haugen and M. Traetteberg, *Acta Chem. Scand.* **20**:1726 (1966).

226. C. A. Coulson, *Proc. Roy. Soc.* **A169**:413 (1939).

227. R. S. Mulliken, *Tetrahedron* **6**:68 (1959).

228. S. Skaarup, J. E. Boggs, and P. N. Skancke, *Tetrahedron* **32**:1179 (1976); B. M. Mikhailov, *Tetrahedron* **21**:1277 (1965); H. J. Bernstein, *Trans. Faraday Soc.* **67**:1649 (1961); H. C. Longuet-Higgins and L. Salem, *Proc. Roy. Soc.* **A251**:172 (1959).

229. H. Kollmar, *J. Am. Chem. Soc.* **101**:4832 (1979).

230. J. P. Daudey, G. Trinquier, J. C. Barthelat, and J.-P. Malrieu, *Tetrahedron* **36**:3399 (1980).

231. G. W. Wheland, *Resonance in Organic Chemistry*, Wiley, New York, 1955.

232. B. P. Stoicheff, *Can. J. Phys.* **32**:339 (1954).

233. J. Kao and N. L. Allinger, *J. Am. Chem. Soc.* **99**:975 (1977); A. Almenningen and O. Bastiansen, *Skr. K. Nor. Vidensk. Selsk.* **4**:1 (1958).

234. S. Fliszár and C. Minichino, *Can. J. Chem.* **65**:2495 (1987).

235. S. Fliszár and C. Minichino, *J. Phys. Colloque C4* **48**:367 (1987); S. Fliszár, in *Chemistry and Physics of Energetic Materials*, S. N. Bulusu (ed.), *NATO ASI Series C* **309** (1990).

236. D. F. McMillen and D. M. Golden, *Ann. Rev. Phys. Chem.* **33**:493 (1982); K. W. Egger and A. T. Cocks, *Helv. Chim. Acta* **56**:1516 (1973); H. E. O'Neal and S. W. Benson, in *Free Radicals*, Vol. 2, J. K. Kochi (ed.), Wiley, New York, 1973, pp. 275–359; D. M. Golden and S. W. Benson, *Chem. Rev.* **69**:125 (1969); G. Leroy, in *Computational Theoretical Organic Chemistry*, I. G. Csizmadia and R. Daudel (eds.), Reidel, Dordrecht, 1981, pp. 253–334.

237. A. L. Castelhano and D. Griller, *J. Am. Chem. Soc.* **104**:3655 (1982).

238. J. C. Schultz, F. A. Houle, and J. L. Beauchamp, *J. Am. Chem. Soc.* **106**:3917 (1984).

239. S. W. Benson, *Thermochemical Kinetics*, 2nd ed., Wiley, New York, 1976.

240. J. C. Traeger and R. G. McLoughlin, *J. Am. Chem. Soc.* **103**:3647 (1981).

241. A. Kerr, *Chem. Rev.* **66**:465 (1966); D. R. Stull and H. Prophet, Natl. Stand. Ref. Data Ser., Bureau of Standards NSRDS-NBS **37** (1971); M. H. Baghal–Vayjoosee, A. J. Colussi, and S. W. Benson, *J. Am. Chem. Soc.* **100**:3210 (1978); *Int. J. Chem. Kinet.* **11**:147 (1979); R. J. Shaw, *J. Phys. Chem. Ref. Data* **7**:1179 (1978).

242. S. W. Benson, F. R. Cruikshank, D. M. Golden, G. R. Haugen, M. E. O'Neal, A. S. Rodgers, R. Shaw, and R. Walsch, *Chem. Rev.* **69**:279 (1969).

243. H. Spiesecke and W. G. Schneider, *J. Chem. Phys.* **35**:722 (1961).

244. R. T. Sanderson, *J. Org. Chem.* **47**:3835 (1982).

245. R. T. Sanderson, *Chemical Bonds and Bond Energy*, 2nd ed., Academic Press, New York, 1972.

246. S. G. Lias, J. E. Bartness, J. F. Liebman, J. L. Holmes, R. D. Levin, and W. G. Mallard, Gas-phase ion and neutral thermochemistry, *J. Phys. Chem. Ref. Data* **17**(Suppl. 1) (1988).

247. NIST Standard Reference Database, *NIST Chemistry WebBook*, 2000.

248. J. D. Cox and G. Pilcher, *Thermochemistry of Organic and Organometallic Compounds*, Academic Press, London, 1970.

249. Y. Hamada, N. Tanaka, Y. Sugawara, A. Hirakawa, M. Tsuboi, S. Kato, and K. Morokuma, *J. Mol. Spectrosc.* **96**:313 (1982).

250. D. K. Dalling and D. M. Grant, *J. Am. Chem. Soc.* **89**:6612 (1967).

251. S. Odiot, M. Blain, E. C. Vauthier, and S. Fliszár, *J. Mol. Struct.* (*Theochem*) **279**:233 (1993).

252. E. L. Cavalieri, E. C. Vauthier, A. Cossé-Barbi, and S. Fliszár, *Theor. Chem. Acc.* **104**:235 (2000).

253. E. L. Cavalieri and E. Rogan, in A. H. Neilson (ed.), *The Handbook of Environmental Chemistry*, Vol. 3J: *PAH and Related Compounds*, Springer-Verlag, Berlin/ Heidelberg/New York, 1998, pp. 81–117.

254. D. R. Lide (ed.), *CRC Handbook of Chemistry and Physics*, 72nd ed., CRC Press, Boston, 1991.

255. F. D. Rossini, *Selected Values of Physical and Thermodynamic Properties of Hydrocarbons and Related Compounds*, Carnegie Press, Pittsburgh, PA, 1952.

256. G. R. Somayajulu and B. J. Zwolinsky, *J. Chem. Soc., Faraday Trans. II* **68**:1971 (1972); *ibid.* **70**:967 (1974).

257. W. D. Good, *J. Chem. Thermodyn.* **2**:237 (1970).

258. C. W. Beckett, K. S. Pitzer, and R. Pitzer, *J. Am. Chem. Soc.* **69**:2488 (1947).

259. D. H. R. Barton, O. Hassel, K. S. Pitzer, and V. Prelog, *Science* **119** (1953).

260. D. K. Dalling, D. M. Grant, and E. G. Paul, *J. Am. Chem. Soc.* **95**:3718 (1973).

261. D. K. Dalling and D. M. Grant, *J. Am. Chem. Soc.* **96**:1827 (1974).

262. C. J. Egan and W. C. Buss, *J. Phys. Chem.* **63**:1887 (1959).

263. E. J. Prosen, W. H. Johnson, and F. D. Rossini, *J. Res. NBS* **37**:51 (1946).

264. R. L. Montgomery, F. D. Rossini, and M. Manson, *J. Chem. Eng. Data* **23**:125 (1978).

265. J. K. Choi, M. J. Joncich, Y. Lambert, P. Deslongchamps, and S. Fliszár, *J. Mol. Struct.* (*Theochem*) **89**:115 (1982).

266. J. B. Pedley and J. Rylance, *Computer Analysed Thermochemical Data: Organic and Organometallic Compounds*, Univ. Sussex, UK, 1977.

267. G. J. Abruscato, P. D. Ellis, and T. T. Tidwell, *J. Chem. Soc., Chem. Commun.* 988 (1972).

268. J. W. de Haan, L. J. M. van der Ven, A. R. N. Wilson, A. E. van der Hout–Lodder, C. Altona, and D. H. Faber, *Org. Magn. Res.* **8**:477 (1976).

269. M. J. S. Dewar and C. de Llano, *J. Am. Chem. Soc.* **91**:789 (1969).

270. C. J. Finder, M. G. Newton, and N. L. Allinger, *Acta Crystallogr. Sect. B* **30**:411(1974); A. Koekstra, P. Meertens, and A. Vos, *Acta Crystallogr. Sect. B* **31**:2813 (1975).

271. M. Traetteberg, E. B. Frantsen, F. C. Mijhoff, and A. Koekstra, *J. Mol. Struct.* **26**:57 (1975).

272. M. Traetteberg and E. B. Frantsen, *J. Mol. Struct.* **26**:69 (1975).

273. O. Bastiansen and M. Traetteberg, *Tetrahedron* **17**:147 (1962).

274. P. R. Pinnock, C. A. Taylor, and H. Lipson, *Acta Crystallogr.* **9**:173 (1956); F. R. Ahmed and J. Trotter, *Acta Crystallogr.* **16**:503 (1963).

275. J. B. Stothers, C. T. Tan, and N. K. Wilson, *Org. Magn. Reson.* **9**:408 (1977).

276. R. Munday and I. O. Sutherland, *J. Chem. Soc. B* 80 (1968).

277. H. L. Finke, J. F. Messerly, S. H. Lee, A. G. Osborn, and D. R. Douslin, *J. Chem. Thermodyn.* **9**:937 (1977).

278. W. D. Good, *J. Chem. Thermodyn.* **5**:715 (1973).

279. D. E. Gray (ed.), *American Institute of Physics Handbook*, 3rd ed., McGraw-Hill, New York, 1972, p. 905.

280. ICSU-CODATA Task Group on Key Values for Thermodynamics, *J. Chem. Thermodyn.* **3**:1 (1971).

281. D. D. Richardson, *J. Phys. C. Sol. State Phys.* **10**:3235 (1977); E. Santos and A. Villagrá, *Phys. Rev. B* **6**:3134 (1972); R. J. Good, L. A. Girifalco, and G. Kraus, *J. Phys. Chem.* **62**:1418 (1958).

282. A. D. Becke, *J. Chem. Phys.* **98**:1372 (1993).

283. V. Barone, *Chem. Phys. Lett.* **226**:392 (1994).

284. V. Barone, *Theor. Chim. Acta* **91**:113 (1995).

285. M. W. Chase, Jr., I. L. Curnutt, J. R. Downey, Jr., R. A. McDonald, A. N. Syverud, and E. A. Valenzuela, *JANAF Thermochemical Tables*, Natl. Stand. Ref. Data Ser. (Suppl. 1), Natl. Bureau of Standards (1982).

286. M. W. Chase, Jr., I. L. Curnutt, J. R. Downey, Jr., R. A. McDonald, A. N. Syverud, and E. A. Valenzuela, *J. Phys. Chem. Ref. Data* **11**:595 (1982).

287. D. W. Scott, *J. Chem. Thermodyn.* **3**:843 (1971).

288. G. Gamer and H. Wolff, *Spectrochim. Acta* **A29**:129 (1973).

289. A. L. Verma, *Spectrochim. Acta* **A27**:2433 (1971).

290. J. N. Gayles, *Spectrochim. Acta* **A23**:1521 (1967).

291. K. Hamada and H. Morishita, *Z. Phys. Chem.* **97**:295 (1975).

292. R. G. Snyder and G. Zerbi, *Spectrochim. Acta* **A23**:391 (1967).

293. J. O. Fenwick, D. Harrop, and A. J. Head, *J. Chem. Thermodyn.* **7**:943 (1975).

294. M. Iborra, J. F. Izquierdo, J. Tejero, and F. Cunill, *J. Chem. Eng. Data* **34**:1 (1989).

295. E. J. Smutny and A. Bondi, *J. Phys. Chem.* **65**:546 (1961).

296. K. Byström and M. Månsson, *Perkin Trans.* **3**:565 (1982).

297. J. K. Crandall and M. A. Centeno, *J. Org. Chem.* **44**:1183 (1970).

298. E. G. Paul and D. M. Grant, *J. Am. Chem. Soc.* **85**:1701 (1963).

299. R. Fuchs and L. A. Peacock, *Can. J. Chem.* **57**:2302 (1979).

INDEX

Alcohols
  atomic charges, 86
    and ^{17}O NMR shifts, 86, 199
  enthalpies of formation, 200
  parameters, 128, 199
  ZPE + heat content, 200
Aldehydes
  atomic charges, 86, 202
    and ^{17}O NMR shifts, 202
  energy formula, 201, 202
  enthalpies of formation, 203
  parameters, 201, 202
  ZPE + heat content, 109, 203
Alkanes
  atomic charges, 54–63, 90
    and ^{13}C NMR shifts, 72–74, 89,
      169, 171
  dissociation energies, 160–163
    charge neutralization, 160–163
  energy formula, 132, 171
  enthalpies of formation, 170, 174
  parameters, 128
  ZPE + ΔH, 104–106, 174
Amines
  atomic charges of carbon, 76, 189, 190

    and ^{13}C NMR shifts, 87
  atomic charges of nitrogen, 6, 62, 190,
    193, 194
    and ^{15}N NMR shifts, 78–80, 87
  carbon–nitrogen bonds, 192, 193
    dissociation of, 164, 165
  electrostatic potential at H,
    191, 192
  enthalpies of formation, 195
  N–H bonds, 191, 192, 194
  parameters, 128
  ZPE + heat content, 108, 109
Ammonia, 109, 191
  electrostatic potential at H, 191
Aromatic hydrocarbons
  atomic charges, 62, 144
    and ^{13}C NMR shifts, 68–72
  parameters, 144–148, 186–188
  enthalpies of formation, 184–188
  ZPE + ΔH, 107, 108, 184–185
Atom
  core–valence separation, 18–33
  in molecules, 4, 39–52
  relativistic corrections, 32
  enthalpies of formation, 103

Atomic Charges, Bond Properties, and Molecular Energies, by Sándor Fliszár
Copyright © 2009 John Wiley & Sons, Inc.

Atomic charges, 6, 7
Del Re approximation, 7, 195
and inductive effects, 53–62
and ionization potentials, 78, 84,
89–91, 95
Mulliken-type analysis, 6, 7, 93–97
and NMR shifts, 59, 65–87, 91,
132, 171
reference charges, 62
Atomization energy, 39, 49, 102, 113
formula for alcohols, 199–201
for alkanes, 132
for amines, 194, 195
for cycloalkanes, 132, 171–174
for carbonyl compounds, 201–203
for ethers, 198, 199
for olefins, 178–181

Benzene, 5, 46, 48, 69, 72, 144,
162, 184
Bond dissociation *see* Dissociation
Bond energy
basic formula, 117, 118, 124
alternative formula, 125
empirical formula, 130, 131
of carbon–carbon bonds, 5, 128
of C–H bonds, 128, 129, 161–163
of C–N bonds, 164, 165
of amines, 129, 165
of nitroalkanes, 165
of glycosyl CN bonds, 165
of carbon–oxygen bonds
of alcohols, 128
of aldehydes, 201
of ethers, 128, 129
of ketones, 201
of O–H bonds, 128, 129, 199–201
of N–H bonds, 128, 129
of N–N bonds, 128, 129
reference values, 128
Butane-*gauche* effects, 171–173
and ZPE + ΔH, 104–106

Carbonyl compounds
atomic charges, 76, 77, 201, 202
and ^{17}O NMR shifts, 84, 85, 202
enthalpies of formation, 203
parameters, 201, 202
ZPE + heat content, 108, 109, 203

C^+–H^- polarity, 58, 61
Charge neutralization energy, 154,
155, 159
Configuration interaction, 12–16,
20–22
Conjugation, 142–144
Core–valence separation, 17
core electrons
energy formula, 30
in molecules, 43–45
exchange integrals K^{cv}, 11, 20–25
separation criteria, 18–22
Politzer–Parr approximation, 18, 19,
33, 51
valence electrons
energy of, 30, 33, 43, 46, 47
Coulomb integrals, 10, 11
Cycloalkanes
atomic charges, 72–74
butane-*gauche* effects, 104,
171–173
enthalpy calculations, 171, 174
ZPE + heat content, 104–106
Cyclopropane, 5, 175

Debye zero-point energy, 108
Dienes, 107, 142–144, 180, 182
Dipole moments, 7, 14
Dissociation energy
of carbon–carbon bonds, 160
of C–H bonds, 161–163
of C–N bonds, 164, 165
of glycosyl CN bonds, 165
of N–N bonds, 194
reorganizational energy, 154, 157
Sanderson formula, 159, 161–163

Electrostatic potential
at H nuclei, 118, 119, 191, 192
Empirical bond energies, 130, 131
Enthalpy of formation
atoms, 103
basic formula of, 101, 102
numerical results, 170, 174, 181–188,
195, 199, 200, 203
Ethers
atomic charges, 76, 85
and ^{13}C NMR shifts, 76
and ^{17}O NMR shifts, 83–86

enthalpy of formation, 199
general formula, 198
parameters, 128, 129
ZPE + heat content, 108–110, 199
Ethylene
carbon charge, 62, 74, 75, 95, 96
bond energy, 5, 141, 144, 147
π orbital centroid, 137–141
Exchange integrals K^{cv}, 11, 20–22

Gaussian bases, 12
Gauss' theorem, 33, 46, 114
Graphite, 108, 187

Heat content, 103
of cycloalkanes, 104–106
Hellmann–Feynman theorem, 4, 27, 37,
114, 117
Hydrazine, 108, 109, 191, 192, 194

Inductive effects
and atomic charges, 54–59
and ionization potentials, 95
and nuclear magnetic resonance
shifts, 59–61
and polar σ^* constants, 54–57
Taft's scale of, 54
Intrinsic bond energy
basic formula, 117, 118, 124, 125
alternative formula, 125
Ionization potentials
of alkanes, 89–91
of amines, 78
of ethers, 83, 84

Ketones
atomic charges, 76, 77, 84, 85
energy formula, 201, 202
enthalpies of formation, 203
ZPE + heat content, 109

Mulliken population analysis, 6, 7, 93, 94
modified analysis, 7, 94–97
overlap population, 93, 94
Multiple carbon–carbon bonds
acetylene, 5
aromatics, 5, 144–147
ethylene, 5, 141, 144
graphite, 187, 188

Net charges *see* atomic charges
Nitrogen NMR shifts
of amines, 78–80
solvent effects, 80
of diazines, 81
of isonitriles, 80
of nitroalkanes, 80
of pyridines, 81, 82
of triazine, 81–83
Nitroalkanes, 164, 165
Nitromethane, 164, 165
Nitrogen–nitrogen bonds, 191, 194
Nonbonded interactions, 113, 115–117,
156, 199, 200, 203
Nuclear magnetic resonance
and atomic charges, 65–68
of sp^2 carbons, 68–72, 74, 75
of sp^3 carbons, 72–74, 76
of nitrogen, 77–83
of oxygen, 83–86
solvent effects, 79, 80, 82, 83

Olefins
atomic charges, 61, 62, 74
and ^{13}C NMR shifts, 74, 75, 95, 96
energy formula, 178–181
parameters, 141, 142, 144, 147, 180
ZPE + heat content, 106, 107

Parameters
alcohols, 128, 199–201
aldehydes, 201–203
alkanes, 128, 132, 147, 169, 171
amines, 128, 192, 193
aromatics, 144–147
ethers, 128, 197, 198
ketones, 201–203
olefins, 144–147, 178–180
Pi-orbital centroids, 134–141
Polar substituent constants, 54–56
Politzer formula, 28
Politzer–Parr formula, 19, 33, 51
Pyridines
N charges and NMR shifts, 81–83
solvent effects, 82

Radicals
atomization energy, 157, 158
ZPE + heat content, 110

Relativistic corrections, 32
Reorganizational energy, 157–159,
    163
Ring strain, 175
Ruedenberg approximation, 28

Sanderson's approximation, *see*
    Dissociation energy
Single–double configuration interaction
    calculations, 12–16
    in charge analyses, 15, 16, 62
    and core-valence exchange integrals,
        21–23
    and valence region energy, 30–32
Slater bases, 12
Solvaton model, 68

Taft's polar constants, 54–56
Thomas–Fermi theory, 19, 28, 114
    approximations of, 19, 33, 114

Valence electrons
    in Hartree–Fock space, 17, 18
    in real space, 20–23

in molecules, 46–49
    real-space energy, 30, 33, 42, 46, 47
Valence orbitals, 17, 18, 134–136
Valence state, 123
Virial theorem, 28, 30, 37, 126

Water, 199–201

Xα calculations, 96, 127, 128

Zero-point energy, 103
ZPE + heat content
    of alcohols, 200
    of aldehydes, 108, 109, 203
    of alkanes, 104, 170
    of amines, 108, 109, 195
    of aromatics, 107, 108, 184, 185
    of cycloalkanes, 104–106, 174
    of ethers, 108, 109, 199
    of graphite, 108, 187, 188
    of ketones, 108, 109, 203
    of olefins, 106, 107
    of radicals, 110
    and butane-*gauche* effects, 104–106